THE SCIENCE BEHIND
TECHNOLOGY

GERCEIDA JONES, PH.D.

Kendall Hunt
publishing company

Kendall Hunt
publishing company

www.kendallhunt.com
Send all inquiries to:
4050 Westmark Drive
Dubuque, IA 52004-1840

Published in the United States of America

CONTENTS

PREFACE

This book follows the intertwined histories of science, technology, and society, focusing mainly on the technology of communication. It elucidates how technological developments are inspired by scientific investigations and these investigations are inspired by inventive technology. Interspersed are the stories of the creative personalities who provided the theories, applied the science, or conceived the inventions as well as studies of the impact technology has had on societies.

Most of the science in this textbook is learned through inquiry-based group activities and student projects rather than straight lecture mode. These activities are a mix of hands-on and computer-based experiments that illustrate the scientific method and the role of experimentation in producing scientific results, while illuminating the science behind the particular technology being studied. The highest level of math required to understand the concepts in this book is Algebra. The mathematical equations used will help to clarify ideas, understand basic concepts, and the experiments performed. Scientific areas include the basic principles of acoustics, electromagnetism, magnetic forces, voltage, capacitors, inductance, the wave and quantum nature of light, and quantum electronics in order to understand the technology of communication. We will start our journey discussing the earliest forms of communication from the drums in Africa to convey message to the evolution of the telegraph and finally cell phones and beyond.

As the chapters progress, other technologies such as the telephone, radio, television, lasers, cameras, communication satellites, and satellites in general used in astronomy will be discussed. We will take a look at future technological possibilities, such as drones, and how they are used in almost every part of daily life. Without the battery most of the products we use would not be possible. We will look at the first modern battery dating back to the 1800's and how new companies are creating even more sophisticated ones for longer use to increase the quality of life for everyone on this planet.

The intent of this book is to provide the reader/student with enough knowledge, skill, and confidence to present a research projects on the concepts learned or just to get a basic understanding of how the universe works to some degree. The projects that students will be able to present after reading this textbook will result in written and oral presentations on the basic science, history, and impact technology has had in one area of either business, energy, entertainment, medicine/nanotechnology, robotics, the digital world, and transportation. Finally, this book will interlace student comments on experiments, simulations, and how to improve activities throughout the textbook. Often I comment to my students that they are instrumental in making this a piece of work that can be used in any college or university.

© Shutterstock.com

CHAPTER 1

Introduction to the Science Behind Technology

What is physics? It is the study of energy, matter and their interaction with one another. The Greek word for physics is "physis" which means "nature". Energy appears in various forms, such as, chemical, electrical, mechanical, nuclear, radiant, thermal and even gravitational fields. It is the fields that carry energy and momentum. Gravity creates potential energy from a high point to a low one. In this course we will eventually discuss each of those concepts. When we look at matter, the standard definition is anything that has mass which can range in size from a subatomic particle to galaxies to the entire universe! However, keep in mind that visible ordinary matter is only a half of percent of atoms and another four percent are invisible in the entire universe while the rest is in the form of dark matter. This unknown mass makes up approximately twenty-three percent of the total mass in the universe. It is invisible and holds together our Milky Way and all other galaxies. Vera Rubin, a graduate student of Vassar College who studied the rotational curves of galaxies, did not discover it until the twentieth century. Unfortunately, this noted scientist passed away on December 25, 2016 as another unsung female hero.

We can start to explain the interaction of energy and matter by looking at the forces that govern them. In a real sense, physics can be seen as the most fundamental natural science of all the sciences but not nearly as complex as biology. It is a tool to guide us in understanding the laws of nature. In ancient times humans did not view the world as we do today. As a matter of fact, supernatural forces dominated their lives. The Greeks are credited with making the first models of nature without relying on myth. However, modern science has come a long way in explaining the fundamental forces that operate in this vast universe. There are four fundamental forces: (1) gravity, (2) the electromagnetism, (3) the weak, and (4) the strong. However, one can argue that dark energy is also a force; a repulsive force that powers the expansion of our universe. It makes up about seventy percent of the density of the universe. Dark energy makes space repel itself. It seems to be a property of space itself. It too is a mystery and not well understood.

1.1 GRAVITY—HOLD ON!

Each of the fundamental forces can be described and applied to our everyday life. Forces supply the means through which particles interact and exchange momentum, and/or attract or repel one another depending on their properties. Particles with mass interact with one another through the force of gravity. For example, it was Isaac Newton who gave us classical mechanics to understand how objects move via forces. This discovery paved the way for the "Industrial Revolution" and the introduction of the steam engine under ordinary motion. With the scientific revolution of satellites in space, it is possible to observe extreme ordinary motion. Now astronomers have found hypervelocity stars that escape the galaxy never returning to the galaxy from which they came: a one-way trip to intergalactic space! These huge luminaries with masses two to five times that of the sun, and surface temperatures well above 18,000° F, move at speeds of 700,000 mph. No one could have imagined that in the 18th century! However, along comes Einstein who took the concept of motion a step further by coming up with relativistic mechanics, which describes motion at the speed of light. Max Planck explained what happens on the subatomic level: a science now known as quantum mechanics where objects on that level act as both a wave and a particle. On a daily basis, gravity plays a major role in our lives and also on large-scale structure, such as galaxies but is ineffective between atoms, due to the fact that its strength is weak compared to electromagnetism on that scale. Gravity declines with distance according to the inverse square law in Newton's version of Kepler's Third Law and grows stronger as an object's mass increases. This is why gravity dominates the structure of the universe on large scales.

1.2 LIGHT MY WAY

The second force, known as the electromagnetic (E-M) force is created through the acceleration of charged particles that interact with one another. Electromagnetism is the only force that affects the interaction of atoms. It can group electrons and nuclei into atoms, atoms into molecules, and molecules into living cells. It governs every aspect of chemistry and biology. In a real sense, it is responsible for the very existence of life on this planet. It too declines with distance like the force of gravity. However, it differs from gravity in a very distinctive way, in that it can be either attractive or repulsive depending on the charges of the particles, while gravity always attracts. The E-M force has a long history of evolution with several scientists making a significant contribution to explaining electricity and magnetism, such as James Clerk Maxwell, Michael Faraday, and Thomas Edison. This is the technology that lights up our cities, powers our appliances, and in some cases, runs our cars. We all know what it feels like to be without power; Hurricane Katrina and Sandy showed us how folks lived during a period in history when electricity did not exist in their world. Understanding how electricity and magnetism work, unleashed the "Electronic Revolution" which has forever changed our way of life. It is almost unbelievable that a little more than three decades ago the general population did not have cell phones (1983) or laptops (1981) at their fingertips! In another twenty-five years from now, it is possible that we will have wallpaper that can change color at the push of a button to change the mood in a room and mirrors that talk back to us like the childhood nursery rhyme in Snow White and the Seven Dwarfs. Yes, you will be talking to a mirror saying, "mirror mirror on the wall" instead of talking to Siri. She will be history! You will wonder to yourself, "How did we ever live without this technology"?

1.3   THE STRONG VERSUS THE WEAK

The third and fourth force will be lumped together in this introduction since their strength is within the nucleus. When Einstein wrote his famous equation $E = mc^2$ in 1905 and the atom was split in the 1930s, scientists for the first time began to understand the forces that light up the heavens. This was the answer to "Why does the Sun/any other star shine"? From this discovery, we got the atomic bomb and the idea that this energy could be harnessed. Therefore, the third force, the weak force acts on extremely short-distance scales—it can only be felt within the atomic nuclei. However, it is actually stronger than gravity. It is responsible for nuclear reactions and has no strength beyond the nucleus: a scale as small as 10^{-18} m. Radioactive decay (fusion) converts hydrogen into helium and is what powers the sun. This excess energy from fusion is the source of energy from the sun. When the solar system formed, all rocks contained tiny amounts of these radioactive substances. Radiometric dating can assist geologists in dating the age of rocks and artifacts. It is based on the fixed decay rate of radioactive isotopes based on its half-life. An unstable nucleus will give off radiation because it wants to be stable. An example of this is uranium decaying into lead, which has a half-life of at least 704 million years to 4.7 billion years depending on the type of uranium to date rocks. Carbon-14 dating can be used to date artifacts less than 50,000 years old. It decays into Nitrogen-14. As soon as living things die, they are unable to produce any more Carbon-14. Plants produce Carbon-14 through photosynthesis. This is how we know the age of a dead tree by counting its rings. People and animals don't produce Carbon-14; they ingest it, allowing scientists to determine their ages as well. Carbon-14 has a half-life of 5,730 years. Anything older, such as fossils and bones, scientists use the surrounding rocks. This method is useful in fields such as anthropology, archeology, geology, paleontology, and a new field known as archaeoastronomy.

The fourth force, the strong force is what binds the nucleus together. We can split an atom, nuclear fission, to create energy or fuse together lighter elements to make heavier elements that in turn create energy that makes star's shine via the proton-proton chain converting hydrogen into helium. The strong force is significant because it binds together quarks in clusters to create the familiar subatomic particles, proton and neutrons in the atom. An interesting fact, its origins come from a property known as color. We see color with our eyes but in actuality it is wavelengths! One of the first concepts that we will study in chapter two is the basic principle of waves and it's anatomy.

1.4   IT'S A WRAP

Now going back to the first force; gravity, it is now described by Einstein's Theory of General Relativity while the other three forces can be described by quantum theory which allows us to decode the very very small world of subatomic particles. Richard Feynman who developed diagrams that describe what happens at high speeds on the subatomic level first discovered quantum Field Theory. This sparked a whole technological revolution in the 20th century. We have the transistor, the laser, and digital technology. Even more exciting is the fact that scientists are able to use quantum theory to unlock the secret of the DNA molecule. As a result, the biotechnological revolution is a direct result of computer technology of DNA sequencing which is what machines, robots, and computers do!

The laws of physics as described above can be described clearly without a lot of math. This course is about discovery and applications of these laws. Sometimes it is possible to avoid the use of mathematics by representing your quantitative relationships via a graph. A well-chosen graph is like a good picture, it is worth a thousand words. Analogies are also an excellent way to present mathematical relationships, i.e. concept between energy and money or the structure of an atom versus the structure of the solar system. Using analogies will help with difficult concepts and in fact have played a very important role in developing physics. This is why astronomy will be an intricate part of this course due to the technological shifts in that field.

Ever hear the old saying, "You teach best what you most need to learn"? Well it is true; teaching can enhance one's basic understanding of science by exposure to mathematics and sharing of ideas. Your appreciation for science will increase by knowing something about how the calculations are carried out just like your senses are heightened when a professional musician plays a familiar song. Music is not just enjoyed by the professionals but by amateurs too once they have studied its structure. Therefore, science should not be limited to those with exceptional mathematical ability. We can all learn. Michael Faraday, the celebrated nineteenth-century British scientist, whom we will discuss in this course, was a self-taught orphan. He was a talented person with limited mathematical abilities, yet he became an outstanding physical scientist. Faraday made contributions to both chemistry and physics. He originated the concept of lines of force to represent electric and magnetic fields. Guess how he did this, using rubber bands as and analogy and so will we. James clerk Maxwell, a Scottish theoretical physicist, known for Maxwell's Equations, admitted that without Faraday's physical intuition, he would not have been able to develop his formal mathematical theory of the electromagnetic nature of light.

The role of science is used to explain observed phenomena. We employ the scientific method to help us better understand the problem. Scientific knowledge must be verifiable. Scientists should be able to test the hypothesis over and over and arrive at the same result, preferably in quantitative terms. Science in general, and physics specifically, was once considered a branch of philosophy known as natural philosophy. Metaphysics is a branch of philosophy that deals with the nature of reality, particularly in a secular rather than a theological sense. Metaphysics means "beyond the physics." Questions as to whether there exist unifying principles in nature that are an age-old problem for most scientists has certainly haunted us. There is a film called "What the Bleep do We Know" that challenges modern science as to whether reality is material or ideal or dependent only upon sensory perceptions. This film touches on the realm of metaphysics because miracles or unexplained phenomena that cannot be repeated. Who's to say that this is entirely incorrect? After all, ninety-five percent of the universe is invisible!

Let us now make a distinction between science and technology, which is the goal of this book. A dictionary definition of technology describes it as the "totality of the means employed to provide objects necessary for human sustenance and comfort." An example of these "means" is applied science. Science is described as a systematized body of knowledge, the details of which can be deduced from a relatively small number of general principles by the application of rational thinking. One may think of labeling a television, computer, automobiles, airplanes, or weapons as devices that can be mass-produced or even as products or components of technology. On the other hand, are Newton's laws of motion, the photoelectric effect, the invariance of the speed of light in a vacuum, and semiconductor physics, all of which can be studied by humans useful to the development of technology? The distinctions between science and technology do not always have sharp or clear boundaries. Lasers and their study, for example, can be regarded as belonging to the domain of either science or technology. Although the emphasis in science is on understanding and in technology on application, understanding and application clearly enhance and facilitate each other. It is for this reason that pragmatically oriented societies are willing to spend large sums of money on scientific endeavors even though it may turn out to be not a sound financial investment.

This course like every other introductory or survey level course, deals with both the physical, and natural sciences. It contains a description of the "scientific method", which is suppose to verify the validity of the techniques for discovering scientific information, theories, or solving problems. The scientific method can be broken down into parts: (1) problem/hypothesis, (2) observations and procedures, (3) data, (4) data analysis; could be in the form of a graph, (5) testing, and (6) possibly the modification of the hypothesis. This is a model procedure, even used in physics. Often one will have a hunch or intuition before the start of an experiment. It is always good science to ask the right questions at the right time and hopefully come up with the right answer to the question. More often than not, dumb luck serendipitously plays a role in unveiling the unknown as in the case of Arno Penza and Robert Wilson, physicists working at Bell Labs in New Jersey who found and received the noble prize in physics for discovering the Cosmic Microwave Background Radiation left over from the Big Bang while building a microwave antennae.

Ultimately, however, despite the ambiguities and inconsistencies of individual scientists in their own ways, the demands of the scientific method enumerated above must be satisfied. A scientific theory that does not agree with experiment must eventually be either modified satisfactorily or discarded. A thorough study of science yields power, both in a large sense and in a very real, detailed, and practical sense leading to new technologies. The resulting power is of great concern because it could be used for self-destructive purposes as in Einstein's discovery of the atomic bomb or for good. Science also gives us a deep insight into and an understanding of how nature works. Our physical universe has an underlying order to it. We can see, feel, and sense its marvelous symmetry and rationale. The material to be presented in the succeeding chapters of this book is designed to help you discover the wonderment of our world, and ultimately the universe. Hopefully, you as the student will get an idea of the beauty, simplicity, harmony, and grandeur of some of the basic laws that govern the universe and drives technology. To use an analogy often quoted in science, "it is time to end the overture and to enjoy the first movement of the scientific symphony!"

© bjul/Shutterstock.com

CHAPTER 2

Basic Principles and Waves

2.1 WHAT IS A MECHANICAL WAVE?

For most of us, we learn about the world around us primarily by seeing and hearing. This information comes in the form of waves. We use electromagnetic waves, in particular visible light, to "see" and sound waves allow us to "hear." The classic definition of a mechanical wave is that it is a disturbance that travels or propagates through a medium, such as, a liquid, gas, or solid with time. Mechanical energy comes from the motion, kinetic energy and from the position, potential energy of objects. It is the gravitational potential energy and elastic potential energies that are important forms of potential energy contributing to the mechanical energy of objects. Mechanical waves are ideal because no transfer of matter happens. A *wave* is a *disturbance* that transfers energy from a *source* to a distant *receiver* without the transfer of matter between the two points. For example, imagine a piece of cork resting on the surface of a smooth pond. If you dip your finger in and out of the pond you will observe a "ripple" spread out from your finger and when that ripple arrives at the cork it will move the cork up and down—it will impart energy to it. In this example, the *source,* your finger, has created a *disturbance* (disturbed the level of the water) and that *disturbance* traveled to a *receiver*—the cork. No water actually traveled from the source to the receiver – only the *disturbance* traveled across the surface of the pond. Thus, energy is transferred from a distant source, your finger, to the cork (receiver of the energy) without water (matter) being transferred between the two points. You may have dipped your finger once, creating a *pulse* or maybe you dipped your finger rhythmically up and down, creating a series of *crests* and *troughs* that constitutes a *continuous wave.*

Some mechanical waves, such as electromagnetic waves do not require a medium. Electromagnetic Waves can exist in a vacuum. They consist of periodic oscillations in electrical and magnetic properties that grow, reach a peak, and then diminish to zero in a periodic fashion. We can demonstrate this by looking at one of the PhET computer programs in a later chapter. Mechanical waves are a local oscillation of the medium. Matter is not transported from one region to another just energy. Mechanical work has to be done on the system in order to *disturb* a region of space from its equilibrium position. Once energy is added to the system, the wave will start to propagate from one region to the other until all the energy is used up or dissipated.

In this chapter we will discuss three types of wave phenomena: First, the medium could be a string or rope under tension. Waves on a string carry energy in just one dimension of space along the direction of the string. By shaking the string or rope with an upward motion, you will see this wiggle travel along the length of the rope (see Figure 2.1). Little bits of string are successively *displaced*—moved from their equilibrium (no disturbance) position—as the disturbance reaches them. Using terminology from water waves, the disturbance has high points called *crests* and low points called *troughs*. In Figure 2.1 the *disturbance* is moving along the string from left to right but the bits of string are displaced up and down (crest/trough), perpendicular to the direction in which the wave is travelling. This type of wave is called a *transverse wave*. Electromagnetic waves, (light), are transverse waves and we will discuss them now and at a future time in more detail.

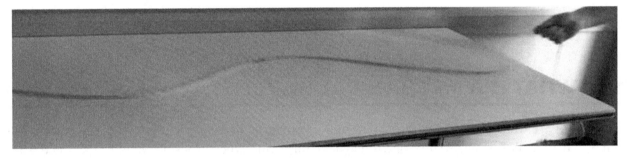

FIGURE 2.1   Wave on a String
Courtesy of Gerceida Jones

The PhET program, "Waves on a String" has three scenarios for you to observe and measure; one end fixed, loose or open end, and no end at all. The displacement of the string or rope as you will observe, is an up/down motion because the direction of travel for the wave is perpendicular along the string. This is known as a transverse wave. Electromagnetic waves (non-mechanical) are of this type and can travel in empty space where there is no medium. Second, if the medium is a liquid or gas as the wave travels horizontally, the displacement of any object (a particle or piece of cork) in the region will move back-and-forth. The displacement of the particles in this medium moves along in the same direction as the wave travels. In other words, the disturbance in this case moves along in the same direction as the wave travels. This is known as a *longitudinal wave and is* characteristic of how sound travels. A longitudinal wave can be observed if you push and pull on a spring as shown in Figure 2.2 (a slinky can be used to demonstrate this concept). Pushing produces, a *compression*—pushing the coils together, makes the coils tighter. Pulling produces a *rarefaction*—pulling the coils apart, making them less dense. The coils move back and forth around their undisturbed positions. Another way to visualize a sound wave is by imagining a speaker that moves in and out, pushing and pulling on air molecules in its surrounding, compressing and rarefying them as indicated in Figure 2.3. As shown in Figure 2.3, the wave is travelling to the right and the molecules are moving back and forth from left to right around their undisturbed positions. Lastly, if the wave is on the surface of a liquid in a channel, such as a canal, the displacement of the water can have characteristics of both longitudinal and transverse motion. In this case if you put a small piece of cork in the water, you will observe as the wave passes by, a circular pattern of the object in the medium occurs as shown by Figure 2.4. Wave interference will be discussed in more detail in a later chapter when we broach the dual nature of light.

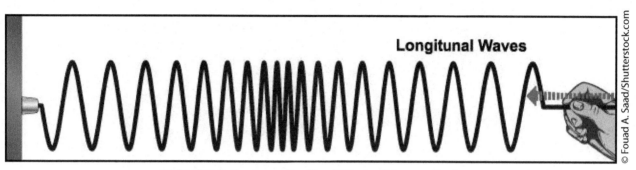

FIGURE 2.2 **Longitudinal Waves Along a Spring**

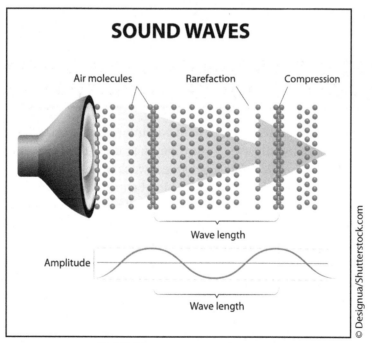

FIGURE 2.3 **Sound Waves**

FIGURE 2.4 Wave Interference

2.2 WHAT IS THE RELATIONSHIP BETWEEN SPEED, FREQUENCY, PERIOD, AND WAVELENGTH FOR ALL WAVES?

What do the three systems described above all have in common? Each example started with the system at *equilibrium* (*rest*), no disturbance initially (see Figure 2.5, the purple line represents at rest—0 amplitude). Energy has to be put into the system by doing mechanical work on it. This disturbance from equilibrium, known as the amplitude (as seen in Figure 2.5—the blue bell shaped curve), will cause a wave motion. This energy will travel from one region of space to another. Before we began our study of waves, first let us describe the anatomy of a wave. You have probably stood on a beach and watched tides as they roll in and out. It appears to have an up and down motion. The top of the wave is called the *crest* and the bottom a *trough* (Figure 2.5 illustration). Note: According to classical mechanics, waves transport energy, but not matter from one region in space to another. The particles do not move. It is the overall wave pattern that carries the energy through the medium. Since the medium does not move, the particles will move either up and down or back-and-forth around their points of equilibrium. Therefore, a wave's anatomy has several parts: (1) crest, (2) trough, and (3) amplitude. The *amplitude*, **A** is the displacement from rest position either the height of the crest above or the dip of the trough below. As the wave propagates it moves through the medium with a certain speed. This is known as the *wave speed*. We can calculate its value by knowing the properties of the medium (air, gas or liquid) in which it travels. The speed at which the wave travels is represented by a *v*. Speed or velocity has SI units of cm/s or m/s). The wave moves with constant speed *v*, the length of one complete wave pattern from crest to crest or trough to trough is known

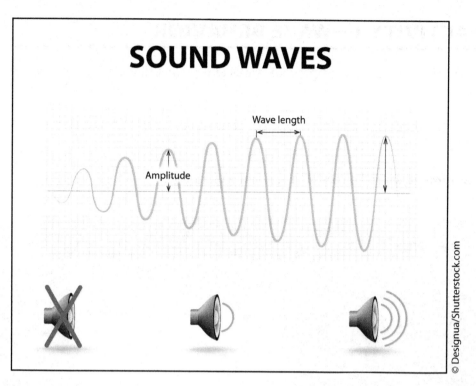

FIGURE 2.5   Shown is the amplitude of the wave from rest position (top-crest) and the (bottom-trough). The distance from crest to crest or trough to trough is one wavelength.

as the *wavelength*, represented with the symbol λ (lambda) with SI units of cm or m. The wave pattern travels with a constant speed v and advances a disturbance of one wavelength in a time interval known as the *frequency*, f. The SI unit for frequency is a cycle per second (1/s), known as a hertz or Hz. Therefore, the wave speed can be defined as $v = f\lambda$. As you can see from this formula, the speed of propagation of the wave is the product of the frequency and wavelength. Note: The frequency is a property of the periodic wave in its entirety because all points on the string must oscillate with the same frequency. You will experiment with periodic waves while simulating the PhET program in Activity 4. Because of the inverse relationship between the frequency and wavelength, if the frequency is increased, the wavelength will decrease so that the integrity of the product, $v = f\lambda$ remains the same and waves of all frequencies propagate at the same speed through the medium.

GROUP ACTIVITY 1—WAVE BEHAVIOR

Students will be able to discuss the property of waves and predict the behavior of waves through various mediums.

Go to http://phet.colorado.edu and download the simulation. The JAVA Development Kit is needed to support the program.

Suggested Equipment: several pieces of string and a giant rope (5–6 feet in length).

Two students demonstrate how to create a sine wave with a large rope.

Choose "Physics" as the science category and find the program "Waves on a String" Click on the link to open the PhET. Play with the program to get use to the features.

Did you notice a difference in the waves as you changed the settings from fixed, loose, and no ends? Are the pieces of string (the balls) moving with the wave across the screen or is just the wave energy moving across the screen?

Turn on the Oscillate button and check fixed end to investigate how waves behave by moving the Amplitude slider from 0–100 and record your observations. How did the waves behave as you moved the slider?

Define amplitude in your own words.

Can you make a connection between what you observed in the program and the rope you have in your hand? If so, explain how.

Now try the same procedure you did with amplitude, only this time vary the frequency and explain your observations.

Try varying the tension, what did you observe? Explain.

Vary the damping mechanism by increasing the number and decreasing it, what happens to the wave?

Set the damping mechanism to 0 and wiggle the wrench to create a single pulse, what did you observe as this pulse traveled to the fixed end and back? Is this scenario ever possible in real life? Record your observations.

How does changing the frequency affect the wave's speed?

How does changing the tension affect the wave's speed?

How does varying the damping mechanism affect the wave's speed?

Change your setting to "loose end" and repeat the steps in the previous experiment.

Try drawing what you observed in this case. Was there a difference between the fixed and loose end scenarios, if so, explain. Was there a part of the wave that appeared not to move?

Change your setting the "no end" feature and set the amplitude on high (frequency, tension and damping mechanism on low) and turn on Oscillate and the Timer buttons. Allow the energy of the string to reach the window, hit the pause button to freeze the wave. (Note: Make sure you reset the timer each time you start and stop the wave). Draw what you observed.

Take a piece of Xerox paper and trace the wave, oscillator, and mark the window from your monitor. Make sure you mark where the green balls are positioned in that picture.

Create a graph with vertical and horizontal positions and label your axes.

Quickly hit the play button and pause. Using that same piece of paper, place it on the monitor, line-up the generator, and trace your new wave. Write down your observations.

Play with this idea some more by changing the settings. Once you have a feel for what is going on and a good picture of your observations, measure the vertical location of the green balls with a ruler (cm). Record the vertical position and the distance by making a graph.

Change your settings back to the "Fixed End" again.

In the green rectangular box, check both "rulers" and "timer". Keeping amplitude set at 0.63, vary the frequency and wavelength while keeping the damping mechanism at 1 tick to the right and tension high. Record 5 trials by creating a Table.

Using the ruler measure the wavelength of your wave created by this scenario. Calculate the wave speed for all 5 trials by using the equation $v = f\lambda$. What is the relationship between frequency and wavelength? Create an Excel graph of wavelength vs. frequency.

Repeat this experiment using a different string tension setting: 2 highs, 2 mediums, and 2 lows summarizing your result. Does tension affect wave speed at all? If so how, show by calculating the wave speed for each trial. Record your data by creating a new Table.

Repeat this experiment one more time by using different amplitudes and summarize your results in Table 2.3. How does amplitude affect the wave speed?

NOTE

It is important to understand waves on a string because there are many applications. Studying transverse waves on a string is a simplified case, as in the study of musical instruments. We will discuss this application in the Sound Lab. The physical quantities that determine the speed of transverse waves on a string are the tension in the string and it's mass per unit length. As you would probably guess, increasing the tension in the string would increase the restoring forces that tend to straighten the string when it is disturbed from a rest position increasing the wave speed. You would also probably guess that if you increase the mass of the string, that would make the motion of the string sluggish and decrease it speed.

1. If you change the amplitude of the wave produced, is there an effect on the wave's speed? If so, explain why you think this is so.

2. If you change the frequency of the wave produced, is there an effect on the wave's speed? If so, explain why you think this is so.

3. Define "frequency" in your own words based upon your observations.

4. If the frequency goes up, what happens to the length of the waves produced? What relationship do frequency and wavelength have?

5. If you change the tension of the string, is there an effect on the wave's speed? If so, explain why you think this is so.

The figure below is to be used for questions 6–9. It shows a rope on a tile floor with a knot at point A. Someone has shaken the end sideways to make a pulse. You are looking down and taking a movie of the motion. Below is one freeze frame of the movie.

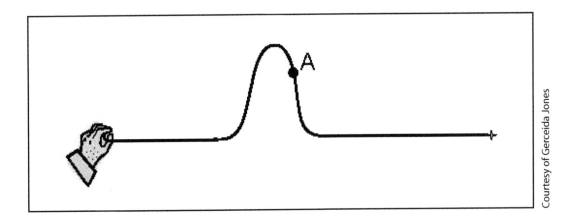

Courtesy of Gerceida Jones

Block out hand and put student hand here

6. If you advance the movie one frame, the knot at point A would be

   a) in the same place

   b) higher

   c) lower

   d) to the right

   e) to the left

7. If the person generates a new pulse like the first but more quickly, the pulse would be

   a) same size

   b) wider

   c) narrower

8. If the person generates another pulse like the first but he moves his hand further, the pulse would be

   a) same size

   b) taller

   c) shorter

9. If the person generates another pulse like the first but the rope is tightened, the pulse will move

   a) at the same rate

   b) faster

   c) slower

The figure below is to be used for questions 10–12. Now the person moves his hand back and forth several times to produce several waves. You freeze the movie and get this snapshot.

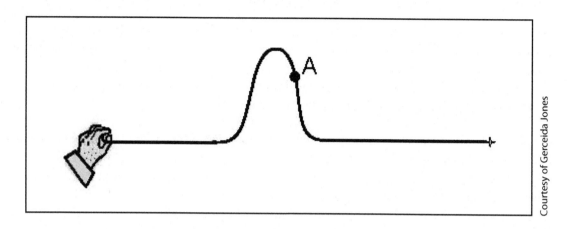

Courtesy of Gerceida Jones

10. If you advance the movie one frame, the knot at point A would be
   a) in the same place
   b) higher
   c) lower
   d) to the right
   e) to the left

11. If you advance the movie one frame, the pattern of the waves will be _____ relative to the hand.
   a) in the same place
   b) shifted right
   c) shifted left
   d) shifted up
   e) shifted down

12. If the person starts over and moves his hand more quickly, the peaks of the waves will be
   a) the same distance apart
   b) further apart
   c) closer together

CHAPTER SUMMARY

The following terminology is used when discussing waves:

▶ **Wavelength:** The *wavelength* of a transverse wave is the distance from one crest to the next crest (or trough to the next trough). Wavelength is traditionally labeled by the Greek letter λ (lambda). The wavelength of a longitudinal wave is the distance from one point of maximum compression (or rarefaction) to the next. From C to C (or R to R) in Figures 2.2 and 2.3. Wavelength is measured in units of length—cm, m, ft, etc.

▶ **Frequency:** If you shake a rope up and down 2 times each second then each successive bit of rope will move up and down 2 times each second. We say the *frequency* of the wave is 2 per second or 2 Hertz—abbreviated 2 Hz. Similarly, if a speaker moves in and out 440 times each second it produces a sound wave of frequency 440 Hz and subsequently the air molecules—or an eardrum—will move in and out with a frequency of 440 Hz. *The frequency of a continuous travelling wave is determined by the source of the wave.*

▶ *Amplitude: Amplitude* determines the energy carried by the wave. For a transverse wave in a rope, for example, the amplitude is the maximum displacement from equilibrium—the maximum height of a crest above equilibrium or the maximum depth of a trough. Large amplitude water waves cause receivers (bits of cork or ships) to have large displacements up and down. For a longitudinal wave it is a measure of the maximum change in particle density. Large amplitude sound waves cause our eardrums to have large displacements and we call such sound "loud."

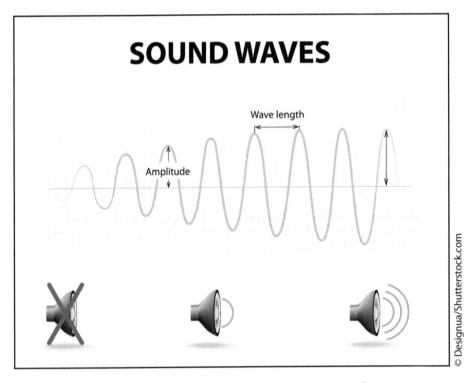

FIGURE 2.6   Amplitude and Wavelength

▶ **Wave Speed:** Recall that the speed of a moving object is the distance travelled divided by the time of travel (speed = distance/time), measured in units of m/s, ft/s, etc. Similarly, for a wave, its speed is the distance a crest (or any other part of the wave) travels divided by time of travel. The speed of a wave is not to be confused with its frequency. Consider the following example: 2 water waves pass by a piece of cork each second (f = 2 Hz) and the distance between crests is 4 meters (λ = 4 m), then 8 meters worth of waves go by each second. Thus the speed of the wave is 8 meters per second (8 m/s). In general, for all kinds of waves, including light waves

$$Wavelength \times frequency = speed$$

But note: *the speed of the wave is determined by the medium through which the wave travels and the conditions of that medium.* Thus, imagine you were making waves in a still pond by dipping your finger in and out. If you "dipped" one time each second you would produce one wave each second or waves with frequency 1 Hz. If you "dipped" 3 times each second you would create a wave with frequency 3 Hz. However, the speed with which the waves spread out is determined by the conditions of the water (how deep, how cold, . . .) and not by the frequency (or the amplitude) of the waves you are producing. Similarly, we will see that electromagnetic waves, including visible light, travel through a vacuum at 300,000 km/s (or 300,000,000 m/s) regardless of their frequency or wavelength and all sound waves travel through air at the same speed.

MATH TO TRY

Example 1:   If the speed of the wave in water is 10 m/s and you dip your finger 2 times each second (f = 2 Hz), then

$$wavelength = \frac{speed}{frequency} =$$

Example 2:   If you now dip your finger 4 times each second (f = 4 Hz) *in the same water,* the speed *is the same* and

$$wavelength = \frac{speed}{frequency} =$$

Wavelength and frequency are *inversely related* so that if the speed remains the same (the medium remains the same) increasing the frequency decreases the wavelength and vice versa. For example, in the example above doubling the frequency results in one half the wavelength of the first example.

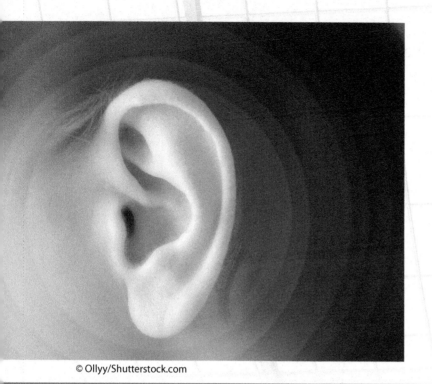

© Ollyy/Shutterstock.com

CHAPTER 3

Sound Waves— We Can Hear

Of all the mechanical waves that occur in nature, the most important in our everyday lives are longitudinal waves in a medium—usually air—called *sound* waves. The reason is that the human ear is tremendously sensitive and can detect sound waves even of very low intensity. In humans the audible range of hearing is approximately 20 Hz to 20,000 Hz depending on the health of the ear. Dogs' hearing depends on its breeding and age; the range is about 40 Hz to 60,000 Hz, much greater than humans. Other animals, such as, bats have very sensitive hearing to compensate for their lack of visual stimuli for navigation. Their range is between 20 Hz to 150,000 Hz while marine mammals hearing range varies from 2,000 Hz to 150,000 Hz. In water, ships use sonar to detect underwater objects while dolphins emit high-frequency sound waves (typically 100,000 Hz) and use the echoes for guidance and for hunting. Whatever the range, hearing is very important for both humans and animals.

When sound waves hit the human ear, it sets the eardrum into oscillation, which in turn causes oscillation of the three tiny bones in the middle ear called the ossicles (Incus, Malleus, and Stapes, see Figure 3.1). These oscillations are finally transmitted to the fluid-filled inner ear and the motion of the fluid disturbs the hair cells (100,000) within the inner ear while simultaneously transmitting nerve impulses to the brain with the information of the sound that is present. Besides their use in spoken communication, our ears allow us to pick up a myriad of cues about our environment, from the welcome sound of your mother ringing the dinner bell to the warning sound of a tornado siren. The ability to hear an unseen nocturnal predator was essential to the survival of our ancestors, so it is no exaggeration to say that we humans owe our existence to our highly evolved sense of hearing. As you've learned in chapter 2, traveling sound waves, transfer energy from one region of space to another that is dependent on *wave intensity*. We will now talk about intensity of a sound wave in terms of the displacement amplitude or pressure amplitude.

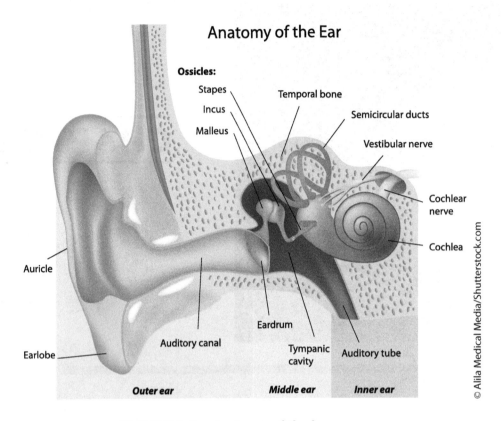

FIGURE 3.1 Anatomy of the human ear

3.1 SOUND WAVES: DISPLACEMENT VS. PRESSURE FLUCTUATIONS

Mechanical waves have been primarily described in terms of displacement. However, a description in terms of pressure fluctuations is often more appropriate because the ear is sensitive to changing pressure within the ear which can be felt every time you fly in an airplane. Let's look at the relationship between displacement, pressure fluctuation, and intensity. When a source of sound or a listener moves through the air, the listener may hear a different frequency from the one emitted by the source. This phenomenon was first described by the 19th-century Austrian scientist Christian Doppler and is called the Doppler Effect, which has important applications in medicine and technology. A familiar sound of the Doppler Effect is the siren of an ambulance, a police car, or a fire truck. If the vehicle is stationary, the whistle sounds the same no matter where you stand (See Figure 3.2). If the object is moving and you are behind it, sound waves stretch to longer wavelengths (lower frequency and pitch—rarefaction) because the object is moving away, you will hear a lower sound. If you are in front of the object, the sound waves bunch up to shorter wavelengths (higher frequency and pitch – compression), and is louder, see Figure 3.3. Just as the vehicle passes by, you hear this dramatic change in pitch from low to high, a sound that sounds a little like a "weeeeeeee-ooooooooooh". Highway Troopers use Doppler to clock your speed as you approach or past him/her using two types of radar; stationary and rolling. Either way, if you're speeding, you're dead meat!

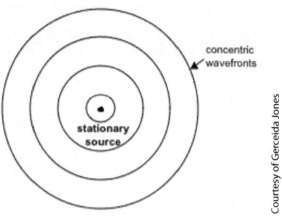

FIGURE 3.2 Stationary Wave Front

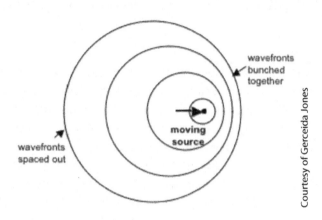

FIGURE 3.3 Moving Wave Front

Medicine uses technology such as *ultrasonic imaging* which operates on the same physical principle; sound waves of very high frequency and very short wavelengths, called ultrasound. The technician uses a wand to scan over the human body, and the "echoes" from interior organs are used to create an image. Ultrasound is used to study heart-value action, detection of tumors, and prenatal examinations. It is more sensitive than x-rays in distinguishing various kinds of tissues and does not have the radiation hazards associated with x-rays.

A sound wave is a longitudinal wave in a medium. Physically it can be observed if you push and pull on a spring as shown in Figure 3.4. Pushing produces a *compression*—pushing the coils together, makes the coils tighter. Pulling produces a *rarefaction*—pulling the coils apart, making them less dense. The coils move back and forth around their undisturbed positions.

Sound can travel through any gas, liquid, or solid. However, most of the sound waves that we encounter on a daily basis propagate in air. In air at normal atmospheric pressure and density, sound travels at 344m/s versus the inner ear sound travels at 1500 m/s. You may be all too familiar with sound traveling through your dorm room walls when your suite mate or next door neighbor is blasting their stereo while you're trying to study. Regardless of where the sound is coming from, we are bombarded by all types of sounds day and night; some pleasant and others annoying either way these ambient noises—diffused waves created by pressure fluctuations

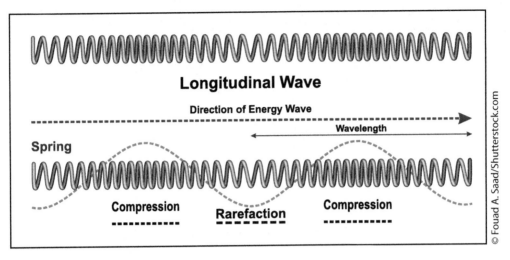

FIGURE 3.4 Longitudinal Waves Along a Spring

in the atmosphere or the scattering of water waves in the oceans can all be traced. Both experimental and theoretical work has shown that diffused waves can produce an elastic response that can be useful in identifying the source from which it came and to determine the properties of the medium. This information is particularly useful in the field of seismology and underwater acoustics. Seismologist use waves to detect earthquakes. By setting up a series of seismometers, researchers can reconstruct the structure and properties of Earth's crust in any part of the world. In the case of ocean acoustics, scientists have used acoustic noise in the ocean to determine the speed of sound and obtain data on the temperature and salinity of the water. Noise can be turned into a signal and have many other applications, such as, interferometry in radio astronomy. We will discuss telescopes in the optics section of this book and how this technology is used to obtain the resolution of a large telescope by assembling multiple smaller telescopes. Further application of ambient noise can be used to observe the surfaces of the Sun and Moon to retrieve helioseismological records. These are just a few ways sound can be converted into signal and used to our benefit. Another example of this very familiar to you is the radio. Sound waves can be digitized too by turning it into ones and zeros. All of these will be discussed at a later time but for now we will primarily focus on sound propagating through air.

3.2 HOW DOES A LISTENER PERCEIVE SOUND?

The physical characteristics of a sound wave are directly related to the perception of that sound by the listener demonstrated by opening the PhET simulation "Sound". For a given frequency, the greater the pressure amplitude of a sinusoidal sound wave, the greater the perceived *loudness* will be. The relationship between pressure amplitude and loudness is not a simple one; it varies from person to person and perceived loudness depends on the health of the ear. A loss of sensitivity at the higher frequencies usually occurs with age. Higher frequency sounds are no longer harmful to older people because the damage has already been done. Sounds that are near our auditory thresholds (under 20 Hz and above 20,000 Hz; normal, healthy hearing range) can't harm us even if they are extremely loud. We don't hear them because they cause no mechanical change in our ear that responds to those frequencies, and consequently no harm is done. In the case of age-related hearing loss, the hair cells in the inner ear that should respond to high-frequency sounds have already stopped. Therefore, there is nothing left to harm. Headphones for personal enjoyment of music, pose a threat to hearing if continuously played at a high volume, like so many riders on NYC or any metropolitan subway.

The frequency of a sound wave is the primary factor in determining the *pitch* of a sound, the quality that lets us classify the sound as "high" or "low". The higher the frequency of a sound within the audible range, the higher the pitch that a listener will perceive, pressure amplitude also plays a role in determining pitch. When a listener compares two sinusoidal sounds waves with the same frequency but different pressure amplitudes, the one with the greater pressure amplitude is usually perceived as louder but also as slightly lower in pitch.

3.3 WAVE INTERFERENCE

Up to this point, we have discussed sound waves that propagate in one direction continuously. What if a wave strikes a shore or a barrier? All parts of the wave are *reflected* back. An example of this occurs whenever you have found yourself in a situation where you are caught between buildings yelling at a friend or in a valley facing a cliff? You hear an echo. The three scenarios that we have observed in the "Waves on a String" simulation mimics what happens in nature. For instance, if you flip the end of a rope that has a fixed end, a pulse travels the length of the rope and is then reflected back to you. Like the echo and reflected wave, the initial and reflected waves overlap in the same region of the medium. This overlapping of waves is known as *interference*. In the case of a transverse wave, what happens when a wave pulse or a sinusoidal wave arrives at the end of the string? If the end is fixed to a support such as a wrench, or wall, the fixed end cannot move. Therefore, the arriving wave exerts a force on the support and the support exerts a force on the string; an equal (string exerts an upward force on the wall) and opposite (the wall exerts a downward reaction force on the string) reaction known as Newton's Third Law. The support's force on the string, sets-up a reflected pulse or wave traveling in the reverse direction. As you will see in the "Sound" PhET simulation, the reflected pulse moves in the opposite direction from the initial (incident), a pulse, and its displacement is also opposite (inverts as it reflects).

Another example we observed; what if the string is attached to a rod with a free end that is able to move in the direction perpendicular to the length of the string? In this case, the string might be tied to a light ring that slides on a frictionless rod perpendicular to the string. The ring and the rod maintain the tension but exert no transverse force. When a wave arrives at the free end, the ring slides along the rod to maximum height, and momentarily comes to rest while the rod exerts no transverse forces on the string. However, the string is now stretched with increased tension so that the free end of the string is pulled back down and a reflected pulse is produced. In this case, the direction of the displacement is the same as for the initial pulse.

The last and third scenario, involves two wave pulses—one right side up and one inverted—traveling in the opposite direction. As the pulses overlap, the displacement of the string at any point is the algebraic sum of the displacements due to the individual pulses. Because the two pulses have the same shape, the total displacement in the middle of the string is zero at all times. In other words, they wash each other out; the meeting of a wave that has a crest and trough of the same amplitude.

3.4 TWO TYPES OF WAVE INTERFERENCE

When two waves arrive at the same place at the same time they "interfere." This interference is "constructive" as when two crests of water waves arrive at the same time and a very large upward movement occurs, or two troughs arrive at the same time and a large downward motion occurs. The interference is "destructive" if a crest and a trough arrive at the same time. If the crest and trough are of equal size the destruction is "total" and the water moves neither up nor down. Interference is a wave phenomenon. In 1801 Thomas Young demonstrated that light exhibits wave properties when he observed interference between two rays of light.

Interference is readily observed for water waves in a ripple tank—shallow pan of water in which vibrating sources produce waves. When light shines through the water and onto a piece of white paper below a glass tank, the crests appear bright and the troughs appear dark. Observe the ripple tank photo in Figure 3.5 which illustrates the interference pattern produced by two coherent sources—sources that vibrate at the same frequency and are "in step" with each other, producing crests (or troughs) at the same time. Note the regions of constructive interference consisting of light and dark bands (crest meets crest—light band, trough meets trough—dark band). Also note the pale radiating lines which indicate regions of destructive interference. Look carefully at these lines and notice that they are places where a crest meets a trough. That is, a dark band on one side of the line meets a light band on the other side. Because the crest and trough "cancel out," at these positions the water level is the same as if there were no waves at all.

THINK ABOUT IT

Referring to Figure 3.5, would you expect a lot of Motion or little motion if a bit of cork was placed on the water's surface at point (1); at (2); at (3); (4); and (5)?

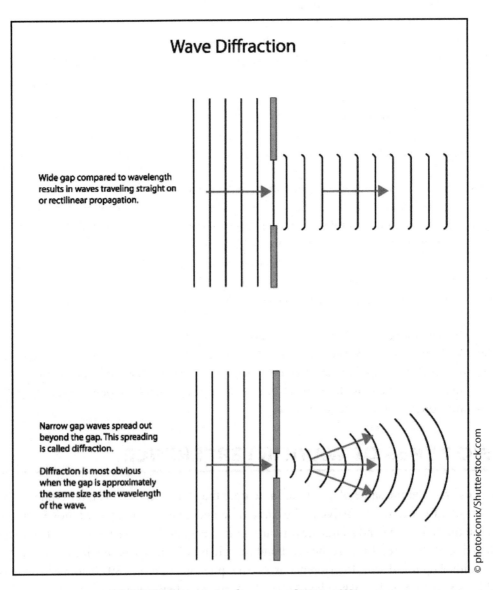

FIGURE 3.5   Interference of Water Waves

3.5   INTERFERENCE OF WAVES—SPEAKERS

Interference of waves occurs when two or more waves overlap in the same region of space as shown in Figure 3.5. A standing wave is a simple example of the interference effect. Here there are two waves traveling in the opposite direction of one another in a medium when combined produce a standing wave pattern with nodes and antinodes that do not move, see the textbox—**Try This** on page 29.

Visualize sound waves by imagining a speaker that moves in and out, pushing and pulling on air molecules, compressing and rarefying them as indicated in Figure 3.6. The figure shows a wave travelling to the right and the molecules are moving back and forth from left to right around their undisturbed positions. There is no inter-

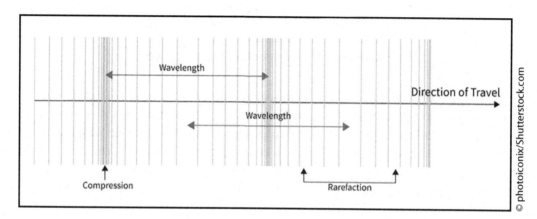

FIGURE 3.6   Sound Waves—Longitudinal

ference. However, if two speakers in phase are driven by the same amplifier, constructive interference occurs as shown by Figure 3.7. The amplifier emits identical sinusoidal sound waves with the same constant frequency and the compressions and the rarefactions of the two waves line up. They strengthen each other and create a new wave with a higher intensity. On the other hand, when the compressions and rarefactions are out of phase, their interaction creates a wave with a dampened or lower intensity. This is called destructive interference,

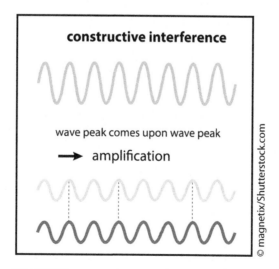

FIGURE 3.7   Constructive Interference

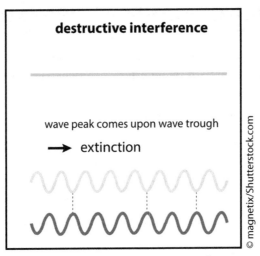

FIGURE 3.8 Destructive Interference

shown in Figure 3.8. In the sound waves simulation, waves that are interfering with each other destructively, the sound is louder in some places and softer in others. As a result, we hear pulses or beats in the sound. In case where two waves cancel each other out; the compressions of one wave line up with the rarefactions from another wave, a dead spot will occur and no sound will be heard. Sound engineers who design theaters or auditoriums must take into account sound wave interference. The shape of the building or stage and the materials used to build it are chosen based on interference patterns. They want every member of the audience to hear loud, clear sounds from their seats.

If we place two speakers side by side, point them in the same direction and play the same frequency, we have constructive interference, see Figure 3.9. Constructive interference occurs whenever the distances traveled by the two waves differ by a whole number of wavelengths; $0, 2\lambda, 3\lambda$; in all of these cases the waves arrive at the listener in phase as demonstrated by the "Sound" program. If the distances from the two speakers to the listener differ by any half-integer number of wavelengths; $\lambda/2, 3\lambda/2, 5\lambda/2$, etc., the waves arrive at the listener out of phase and there will be destructive interference as shown in Figure 3.10. In this case, little or no sound energy will flow towards the listener directly in front of the speaker.

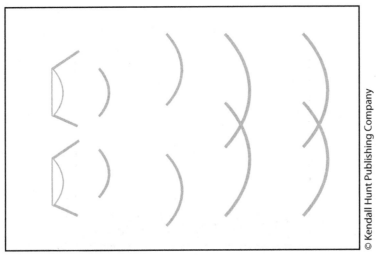

FIGURE 3.9 Two Speakers Side by Side Produces Constructive Interference

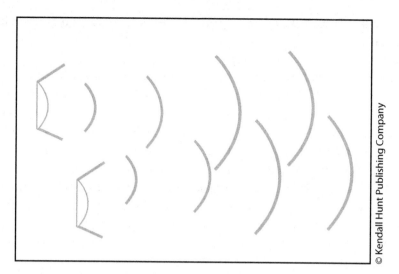

FIGURE 3.10   **Two Speakers One in Front of the Other Produces Destructive Interference**

The experiments shown above are closely analogous to how light behaves too. It provides strong evidence for light's dual nature; it can act as both a particle and a wave. The activities designed for experiments 3 & 4 will demonstrate this concept. Interference effects are used to control noise from very loud sound sources, such as, gas-turbine power plants or jet engine test cells. "The idea is to use additional sound sources that in some regions of space interfere destructively with the unwanted sound and cancel it out. Microphones in a controlled area will feed back signals to the sound sources, which are continuously adjusted for optimum cancellation of noise in the controlled area" (Young and Freedman, 2008).

TRY THIS!

You can create a standing wave in the PhET sims from Lab #1 by setting Amplitude on 20, frequency = 30, Damping to 0, and Tension 1 tick from the right.

3.6   EARLY FORMS OF COMMUNICATION

Africans have used drums to convey messages for centuries. The pitch of the drum closely mimicked the tonal pattern of speech. These messages could be transmitted through dense forest at the speed of a hundred miles in just one hour from village to village. Almost everyone could understand the message but few were able to use the drum as a communication tool. When Europeans came to Africa on expeditions to explore the forest, they would hear the drum beats and described the Africans as "primitive". Little did they know that the Africans had developed a superior form of communication.

The use of the talking drum wasn't noticed by European until the first half of the 18th century. They were surprised that the Africans knew of their coming and intentions long before their arrival. It was Roger T. Clarke, a missionary that realized that the drums were conveying phrases of a poetic nature. Another missionary, John F. Carrington at the age of twenty-four, discovered that the Africans were actually relaying complex messages with their drums. He made Africa his home and learned to drum. In 1949 he published his findings in a book titled "The Talking Drums of Africa".

Other forms of early communication ranged from bonfires on hilltops to setting up a series of weigh stations for carriers on horseback. Over the centuries people tried all kinds of ways to communicate, such as flags, horns, smoke and even mirrors. Through trial and error of with magnets, finally, Samuel F. B. Morse, an American portrait painter would design a pulse system that would travel along a single wire based on European telegraphs. He developed the Morse Code along with his partner, Charles Thomas Jackson. Jackson was well educated in the use of electromagnets. The initial problem with the telegraph was that the signal would not transmit more than a few hundred yards. His break-through came in a conversation with Leonard Gale, a chemistry professor at New York University. With financial backing from a congressman from Maine, Francis O. J. Smith and Congress, he constructed 38 mile experimental telegraph line between Baltimore, MD and Washington D.C. On May 24, 1844, he tapped out this famous line "What hath God wrought!" Morse received a patent for his telegraph in 1847.

The path to long distance communication had been solved. From that point on history tells the story of how technology evolved. James Glick's story, "The Information" is an excellent read for an open discussion on early forms of communication. After reading his article, have the students try the exercise below:

TRY THIS!

Make a list of all the intermediate stages of communicating from the talking drum to cell phones.

3.7 SOUND TECHNOLOGY

Ocean Acoustic Tomography

Although sound in air is very important in our everyday life, so it is with water too. Both humans and animals can use sound to communicate in this medium. As an undergraduate at the University of Michigan studying physical oceanography it became apparent just how useful sound traveling in seawater could be to advancing technology. Listening to whales communicate with each other in class at low frequencies and visually seeing the sine waves move across the oscilloscope was a new and exciting experience. An interesting fact is that the oceans are transparent to low frequency acoustics and the whales were in their natural habitat doing just that!

Sound in seawater travels approximately 1500 m/s depending on the temperature and salinity of the water. The speed of sound can be measured by the time it takes to travel from a known source and to a receiver's location. A good question to ask myself during that experience with the whales was whether acoustics could make the basis for mapping the ocean. The answer to that question is yes. Just as seismologist use P & S waves to get information about the inaccessible interior of Earth, oceanographers could apply this same technology by using acoustic travel times to derive information about this medium.

Since the late 1980's ocean acoustic tomography has evolved. Tomography has been employed to measure a wide range of time scales from a minute, the internal wave field to decades gathering information on climate change. Other applications of tomography are the measurement of ocean tides, modeling ocean basins, and oceanic circulation.

Digitizing Sound

Most people don't understand what digitized sound is all about. If you ask them what does that mean, the response is, "I don't know". But we use this technology on a regular basis from CD's to MP3 players to cell phones to computer games. A simple question to ask is what does the word "binary" mean to you? The most likely answer is it means two. Then I ask what technology uses the binary system, and most students will say computers. In fact, that is exactly how computers operate with only two numbers; 0 and 1. Therefore, sound waves can be converted into binary numbers and then burned onto a CD!

These two numbers 0 and 1 have ancient roots, especially the number 0. It was not just used as a place holder as far back as the 7th century when the Indian astronomer Bhramagupta offered it as a treatment of negative numbers. He actually understood zero to be a number. Its origin would have a say in the development of our modern decimal system, the beginning of modern mathematics. The binary system can be taught. Take the word "Music", it can be stored on a computer as numbers. By using the ASCII table, a standard system that codes each letter and symbol into a number, each piece of information can be digitized. It's not quite this easy but this is how computers store information into binary code.

How do sound waves behave in a digital format? Digital music does not record every single piece of the sound wave instead what it does is sample the wave at set time intervals. The sample (sound wave) is then converted into a sequence of binary numbers. This is how our modern technology that we use at home, work, or play is turned into a sequence of zeroes and ones!

Drones

The history of unmanned aerial vehicles (UAVs) commonly known as drones dates back to the early 1900's. Two inventors Elmer Sperry and Peter Hewitt constructed the radio controlled "flying bomb". Since 2002, the U.S. military has used drones to spy and in combat. Drones also have a commercial use. Technology has made them lighter, cheaper, and more sophiscated. UAVs deliver our packages, take photographs, provide wireless internet, and monitor conservation efforts in remote locations of the world. More recently inventors are finding ways to use UAVs to fight fires! How does this technology work?

A Canadian inventor by the name of Charles Bombardier created Firesound, a type of flying saucer drone that will extinguish small fires with sound waves. Its design is powered by hydrogen fuel cells and lifted by four electric turbofans. Sound can be used to separate burning material from oxygen, extinguishing the fire. This is what Bombardier had to say about his technology: "We share our ideas openly online so they can grow, evolve and inspire people," Bombardier said. "We always seek out new innovative ideas and we encourage young creative entrepreneurs to contact us."

Where is sound technology headed in the future? Who knows! One thing for sure, if sound can be used to fight forest fires maybe, just maybe the hood over your kitchen stove will be able to prevent house fires!

GROUP ACTIVITY 2—SOUND WAVES

Go to http://phet.colorado.edu, click on "Physics" and open the "Sound" simulation.

Click the tab "Listen to a Single Source."

A. Amplitude, Frequency and Wavelength

1. With the audio off, observe how the waves change as you separately adjust the amplitude and frequency and describe your observations by varying the amplitude and frequency.

2. Now enable the audio and listen to how the sound changes as you vary the amplitude and the frequency and describe what you hear.

3. Summarize the connections you can make between what you see in the simulation and what you hear.

4. Move the listener towards and away from the speaker and describe any changes in what you see and what you hear.

B. Speed of a sound wave

The speed of the wave can be measured in two ways—using the distance and time of travel or using the relationship between frequency, wavelength and speed.

Click on the tab that says "Measure". Notice there is a moveable measuring stick that is 5m long and a stopwatch. To determine the speed, you will measure the time it takes for a particular crest or trough to travel 5 m; run a few trials.

speed = wavelength × frequency.

Carefully measure the wavelength and determine the speed by this method. Indicate whether or not you think the speeds determined by these two methods are in "reasonable agreement" and why you think they are or are not.

Change the frequency. What happened? Organize and record your data.

Change the amplitude. What happened? *Organize and record your data.*

Summarize your results for the speed of the wave and indicate any conclusions or comments

C. Interference

When two waves arrive at the same place at the same time they "interfere." This interference is *constructive* as when two crests of water waves arrive at the same time and a very large upward movement occurs, or two troughs arrive at the same and a large downward movement occurs. The interference is *destructive* if a crest and a trough arrive at the same time. If the crest and trough are of equal size the destruction is "total" and the water moves neither up nor down. Similarly, for sound waves, two compressions (rarefactions) result in a region of very compressed (rarefied) molecules but a compression and rarefaction "cancel out." In a later lab we will see that light behaves as a wave because it, too, exhibits interference.

Click the two source interference tab. Look carefully at the dark compressions, white rarefactions and in particular the gray regions. Describe what you notice about the gray regions.

Vary the frequency and describe (in words or pictures) your observations

Enable the audio and move the listener to various positions and describe your observations.

D. What we hear

The compressions and rarefactions of sound waves are regions of higher than normal and lower than normal air pressure, respectively. When a compression reaches our ear, its higher pressure pushes our ear drum in and when the rarefaction arrives, the higher pressure within our ear, pushes the drum out. As a sound wave comes by, our eardrum moves in and out, vibrates, with the frequency of the wave.

Click on "listen with Varying Air Pressure". Enable the audio and describe how the loudness changes as the frequency is varied.

Remove the air and describe what happens visually and audibly while the air pressure is decreasing and when it is removed. Why do you think these changes are occurring?

QUESTIONS AND ANSWERS

1. What is the relationship between musical pitch and the frequency of sound Waves?

2. How do we hear?

3. What frequency range is audible to humans and what effect does our Modern technological society have on that range?

4. How can you change the speed of sound?

5. Doing the wave" at a sports stadium is an example of a mechanical wave. The disturbance propagates through the crowd, but there is not transport of matter from one spectator to another. What type of wave is "the wave"?
   a. longitudinal
   b. combo of transverse and longitudinal
   c. transverse

6. Sound waves are longitudinal waves in air. The speed of sound depends on the temperature; at 20°C it is 344 m/s or 1130 ft/s. What is the wavelength of a sound wave in air at 20°C if the frequency is 262 Hz (middle C on a piano)?

7. An echo is sound reflected from a distant object, such as a wall or a cliff. Explain how you can determine how far away the object is by timing the echo.

8. To increase the volume of a tone at 400 Hz heard by the listener, the speaker must oscillate back and forth more times each second than it does to produce the tone with lower volume.
   a. True
   b. False

9. Which has a more direct influence on the loudness of a sound wave: the displacement amplitude or the pressure amplitude? Explain your reasoning.

10. Two vibrating tuning forks have identical frequencies, but one is stationary and the other is mounted at the rim of a rotating platform. What does a listener hear? Explain.

11. When sound waves travel from air into water, does the frequency of the wave change? The speed? The wavelength? Explain.

12. A sound source and a listener are both at rest on the earth, but a strong wind is blowing from the source toward the listener. Is there a Doppler effect? Why or why not?

Draw a diagram to solve these next two problems (13 & 14).

5. Two loudspeakers, A and B, are driven by the same amplifier and emit sinusoidal waves in phase. The frequency of the waves emitted by each speaker is 172 Hz. You are 8.0 m from A. What is the closest you can be to B and be at a point of destructive interference?

6. Two loudspeakers, A and B, are driven by the same amplifier and emit sinusoidal waves in phase. The frequency of the waves emitted by each speaker is 860 Hz. Point P is 12.0 m from A and 13.4 m from B. Is the interference at P constructive or destructive? Give the reasoning behind your answer.

CHAPTER SUMMARY

The following terminology is used when discussing sound waves:

▶ **Density** of a material is defined as its mass per unit volume.

▶ **Displacement amplitude** is the amplitude A of maximum displacement of a particle in the medium from its equilibrium position.

▶ **Doppler Effect** is the change in the observed frequency of a wave, such as sound or light, occurring when the source and the observer are in motion relative to each other, with the frequency increasing when the source and observer are approaching each other and frequency decreasing when the source and observer are moving away from one another.

▶ **Intensity** is the average rate of flow of sound energy.

▶ **Interference** is an effect that occurs when two or more waves overlap.

▶ **Longitudinal waves** are where the particles of the medium oscillate about their mean positions in the direction of propagation of the wave.

▶ **Loudness** is the perception of how intense sound appears to be; for a given frequency, the greater the pressure amplitude of a sinusoidal sound wave, the greater the perceived loudness.

▶ **Pitch** is the quality of sound primarily determined by the frequency. It lets us classify the sound as "high" or "low". The higher the frequency of sound, the higher the pitch the listener will perceive. Pressure amplitude also plays a role in determining pitch.

▶ **Pressure amplitude** is directly proportional to the displacement amplitude. It depends on wavelength. Waves of shorter wavelength have greater pressure variation for a given amplitude because the maxima and minima are squeezed closer together.

▶ **Pressure fluctuation** is a sinusoidal sound wave in air, the pressure fluctuates above and below atmospheric pressure. It depends on the difference between the displacement at neighboring points in the medium.

▶ **Standard atmospheric pressure** is a unit of pressure; force per unit area exerted on a surface by the weight of air above that surface, units of 101.325 kPa "kiloPascals") or 1013.25 millibars. It is equivalent to 760 mmHg.

© AVKost/Shutterstock.com

CHAPTER 4

Diffraction—Light Waves

In chapter 3 we started the discussion about interference by looking at sound waves, now we want to extend the discussion to light waves. As you know, interference happens when two waves combine. An interesting example that you may have observed is the colors seen in oil films or soap bubbles. This is a result of interference between light being reflected from the front to the back of surfaces in the solution. Effects such as these can be categorized as diffraction. Diffraction can occur whenever a wave passes through an aperture or around an obstacle. They are important in practical applications of physical optics such as diffraction gratings, x-ray diffraction, and holography.

Everyone knows that sound bends around corners or otherwise you would not hear a person in the other room talking to you or the television. However, did you know light can bend? This surprised early 19th century scientists. When light from a point source falls on a straightedge and casts a shadow, the edge of the shadow is never perfectly sharp. Some light appears in the area that we expect to be in the shadow, and we also find alternating bright and dark fringes in the illuminated area as well. The reason for this is that light, like sound, has wave like characteristics. A diffraction grating is a slide of transparent material that allows the wavelength of light to pass through a slit; each infinitesimal part of the slit acts as a source of waves; and the pattern of light and dark bands is a result of interference among the waves emanating from the source.

4.1 HOW DO LIGHT WAVES BEHAVE?

We will start with the interference of coherent sources. In chapter 3 we discussed interference as any two or more waves that overlap in space. When this occurs, the total wave at any point at any instant of time is governed by the principle of superposition as described by the "Waves on a String" PhET simulation. We can apply this same principle to electromagnetic waves. The principle of superposition states: When two or more waves overlap, the resultant displacement at any point and at any instant is found by adding the instantaneous displacements that would be produced at the point by the individual waves if each were present alone.

For transverse waves on a string and longitudinal waves, i.e., sound waves, in both of these cases, waves propagate along a single axis only. However, light waves can travel in 2D or 3D mediums. Now we will see what happens when combined waves spread out in two or three dimensions from a pair of identical wave sources. First we will start with **monochromatic** light (light of a single color). Light bulbs and even a candle flame emit a continuous distribution of wavelengths. However, there is a way to produce an approximate monochromatic light source by using a filter that has a very narrow range of wavelengths such as a laser. Your instructor will most likely demonstrate how a helium-neon laser, which emits red light at 632.8 nm (1 nm = 1.0×10^{-9} m). Figure 4.1 shows Thomas Young's experiment with light passing through a single slit labeled **S1** (a) in the first part of the frozen frame. The single source **S1** shows only the wave fronts corresponding to the wave crests so that each successive wave front is spaced one wavelength apart. Notice that the material surrounding **S1** is uniform, which means the wave speed, is the same in all directions. The long arrows represents the spherical wave fronts moving outward from the source reaching **S2** in phase at a wave speed of $v = \lambda f$. Why do they do this? Because the waves have traveled equal distances from the source.

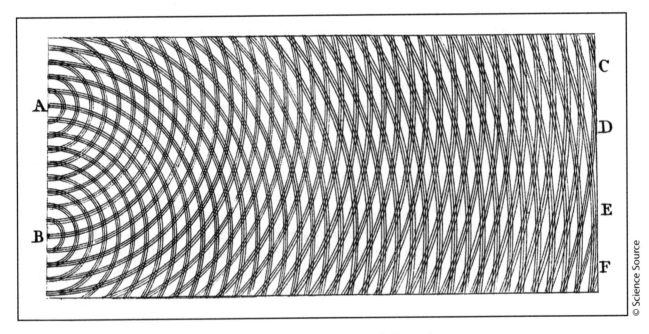

FIGURE 4.1 **Thomas Young's Experiment**

In 1801 Thomas Young created two coherent sources as follows: He used a single light source and split it into two by passing it through two closely spaced narrow slits as seen in the second part of the frozen frame of Figure 4.1 labeled **S2**. Here you have two identical sources of monochromatic waves, **S2** (b & c). The two sources produce waves of the same amplitude **A**, and the same wavelength, λ. Also, the two sources are permanently in phase which means they vibrate in unison. In our everyday life we could think of these sources as two loud-speakers (lab 2) driven by the same amplifier, two radio antennas powered by the same transmitter, or two small holes or slits in an opaque screen, illuminated by the same monochromatic light source. If these items were not in a constant phase relationship, the result would not be the pattern observed on the final screen labeled **F** of Figure 4.1. Two monochromatic sources of the same frequency and with a constant phase relationship are said to be **coherent**.

If the waves emitted by the two coherent sources are transverse, as in electromagnetic waves, then you can assume that the wave disturbances produced by both sources have the same **polarization** (they lie along the same line; i.e., sunglasses). For example, **S2** (b & c) could be two radio antennas in the form of long rods oriented parallel to the z-axis (perpendicular to the plane of the figure); at any point in the xy-plane the waves produced by both antennas have electric fields with only a z-component. Figure 4.2 shows the anatomy of an electromagnetic wave. Then we need only a single scalar function to describe each wave; this makes the analysis a lot simpler.

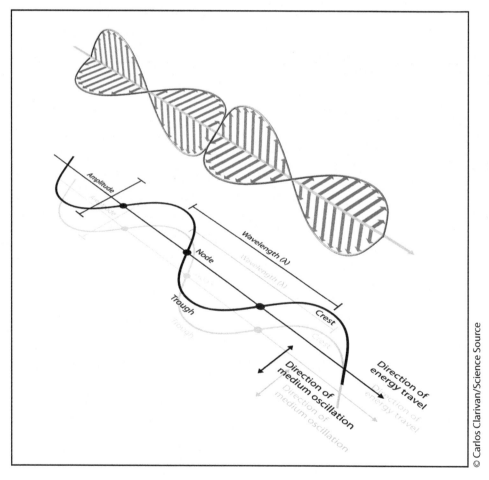

FIGURE 4.2 **Anatomy of an Electromagnetic Wave**

4.2 UNDERSTANDING AN INTERFERENCE PATTERN MORE FULLY

The interference pattern is illustrated in more detail in the drawing of Figure 4.3. The solid lines represent crests and the dashed lines represent troughs and the wavelength is 1 cm. The solid dots are points of maximum constructive interference, either crest meeting crest or trough meeting trough. The open dots are points of total destructive interference, crest-meeting trough (observe this figure carefully).

Note the resulting pattern of alternating regions of constructive and destructive interference. Each of the labeled points is a measurable distance from each of the sources. This distance can be measured in cm or by directly counting wavelengths. For example:

► Point 1 is 4 wavelengths from source *A* and 4 wavelengths from source *B*. Thus, point 1 is *equidistant* from *A* and *B*. In fact, all the points along the line *OD* are points of constructive interference and all are equidistant from sources *A* and *B*. (*Check it out.*)
► The point 2 is 2 ½ wavelengths from source *A* and 3 wavelengths from source *B*. Waves from *B* have traveled an extra 1/2 wavelength as compared to the waves from *A*. In fact, for all the points along the line towards *E* waves from *B* have traveled an extra 1/2 wavelength as compared to the waves from *A*—and these are all points of destructive interference. (*Check it out.*)

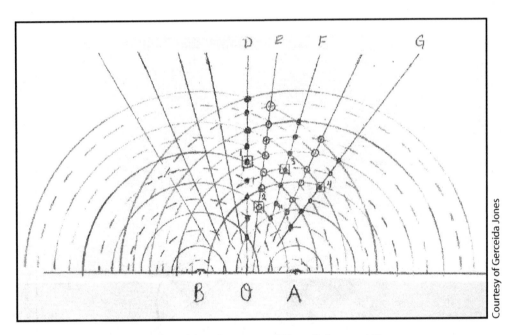

Courtesy of Gerceida Jones

FIGURE 4.3   **Interference of Two Coherent Sources**

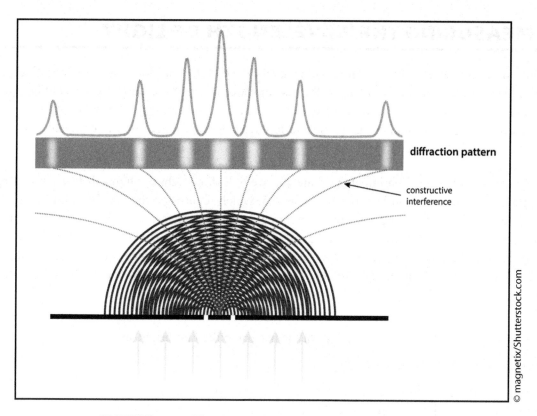

FIGURE 4.4 Thomas Young's Double Slit Experiment

▶ The point 3 is 3 ½ wavelengths from source *A* and 4 ½ wavelengths from source *B*. Waves from *B* have traveled an extra whole wavelength as compared to the waves from *A*. In fact, for all the points along the line towards *F* waves from *B* have traveled an extra whole wavelength as compared to the waves from *A*—and these are all points of constructive interference. (*Check it out.*)

▶ The point 4 is 3 wavelengths from source *A* and 5 wavelengths from source *B*. Waves from *B* have traveled an extra 2 whole wavelength as compared to the waves from *A*. In fact, for all the points along the line towards *G* waves from *B* have traveled an extra 2 wavelengths as compared to the waves from *A*—and these are all points of constructive interference. (*Check it out.*)

This information will allow us to measure the wavelength of light! To visualize the interference pattern, a screen is placed so that the light passing from the slits falls on it in Figure 4.4. The screen will be most brightly illuminated at point A (wave fronts arrive in phase), where the light waves from the slits interfere constructively, and will be darkest at points B (wave fronts arrive out of phase) where the interference is destructive. To simplify the analysis of Young's experiment, we assume that the distance *L* from the slits to the screen is so large in comparison to the distance *d* between the slits that the lines emanating from the source to point A are very nearly parallel. This is usually the case for experiments with light; the slit separation is typically a few millimeters and is given on the diffraction grating slide by the manufacturer while the screen may be anywhere from a meter to many meters away.

4.3   MEASURING THE WAVELENGTH OF LIGHT

Wavelength, often abbreviated by the Greek letter lambda—λ, is the distance between repeating parts of a wave, say, crest to crest or trough to trough. In the case of water waves, for example, you might imagine taking a ruler and measuring the distance from one crest to another. But we cannot see the crests and troughs of light waves. However, using what we know about the interference pattern produced by two coherent sources we will be able to determine the wavelength of light.

As in Young's double slit experiment in Figure 4.4, we will shine light through a diffraction grating onto a distance screen. This arrangement is drawn in Figure 4.5, which focuses on the line OF from Figure 4.3 when F is very far from O. F is the first point of maximum constructive interference—the center of the first *bright spot*, from the center of the pattern D. Recall that all the points along the line OF waves from B have traveled an extra whole wavelength as compared to the waves from A. This extra distance is the distance from B to C in Figure 4.5 (Lines CF and AF are of equal length, so BC is the extra distance). Note the following in this figure:

1. $BC = \lambda$

2. $DF = X_1$ is the distance from the center of the pattern to the first bright spot

3. $AB = d$ is the distance between the slits

4.   OF is the distance from the slits to the first bright spot and for a very distant screen this is essentially the distance from the slits to the screen; $OF = L$

A careful study of this figure reveals that there are two similar triangles. The smaller triangle ABC is a scaled-down version of the larger triangle OFD. That means, if line AB, (the hypotenuse of the smaller triangle) is one tenth as long as line OF (the hypotenuse of the larger triangle), then BC (the shorter leg) is one tenth as long as DF. In general, the ratio of the short legs is the same as the ratio of the hypotenuses.

$$\frac{\lambda}{X_i} = \frac{d}{L} \qquad\qquad \textbf{Equation 4.1}$$

Thus by measuring X_1, d, and L, we can determine λ, the wavelength of light.

$$\lambda = X_i \frac{d}{L} \qquad\qquad \textbf{Equation 4.2}$$

We could have done a similar analysis drawing the line OG where G is the second bright spot from the center, recalling that its extra distance from A is equal to two wavelengths. This analysis would lead to:

$$2\lambda = X_2 \frac{d}{L} \qquad\qquad \textbf{Equation 4.3}$$

where X_2 is the distance from the center of the pattern to the second bright spot. Similar equations can be derived from the 3rd, & 4th, . . . bright spot.

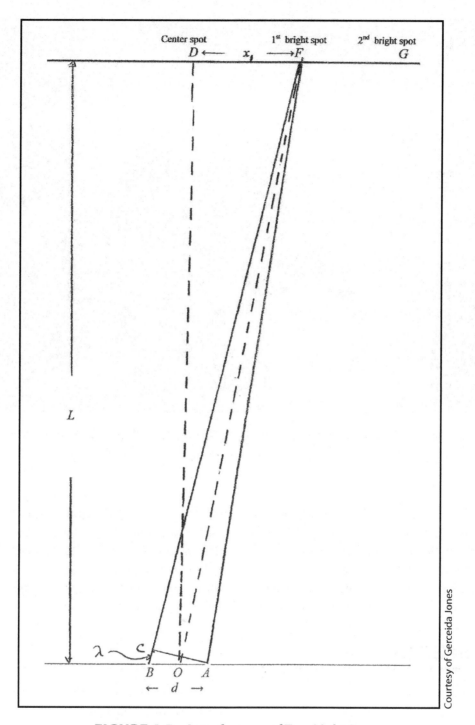

FIGURE 4.5   Interference of Two Light Rays

FIGURE 4.6 Measuring the Wavelength of Light: Students CCW (Daniel, Camille, Kate and Ben)

Students in the Science of Technology course at New York University set-up the experiment to measure the wavelength of light as shown in Figure 4.6. These are just a few of the comments that students made about their experience with the equipment that was provided and the overall difficulty of the task: 1) "It was difficult to stabilize the poles", 2) "Measuring the distance between bright spots on the screen was tedious because the dots kept jumping", and 3) "Getting the laser lined up with the slits in the diffraction grating was difficult", were just a few of the problems in measuring the wavelength of light with this set up.

4.4 DIFFRACTION

We have just finished studying interference patterns that can arise when two light waves are combined. When many light waves are combined, these effects are referred to as diffraction. Actually there is no fundamental difference between interference and diffraction. In this chapter we used interference for effects involving waves from a small number of sources, only two. Diffraction usually involves a continuous distribution of Huygens's wavelets across the area of an aperture, or a very large number of sources or apertures. Both categories of phenomena are governed by the same basic physics of superposition and Huygens's principle. Diffraction is sometimes described as the "bending of light" around an obstacle. But the process that causes diffraction is present in the propagation of every wave. When some obstacle cuts off part of the wave, we observe diffraction effects that result from interference of the remaining parts of wave fronts. Optical instruments typically use only a limited portion of a wave; for example, a telescope uses only the part of a wave that is admitted by its objective lens or mirror. Thus diffraction plays a role in nearly all-optical phenomena.

4.5 ASTRONOMY

Hans Lippershey, a Dutchman in 1608, invented the first telescope. He took two lenses and placed them at each end of a long tube to magnify distant objects. However, in 1609 Galileo Galilei an Italian scientist turned Lippershey's toy instrument into an astronomical tool. Although Galileo's telescopes were low-power refractors, he was able to observe the sun. He soon discovered that the universe was a lot larger than he imagined. Galileo's other observations proved that circles and spheres did not dominate the heavens, the moon had craters and was not a perfect heavenly body, Jupiter's moons, and the sun had spots. His observation that Venus went through phases was the proof that earth was not the center of the universe. Thereby it was a key piece of evidence for the heliocentric model. Telescopes over the past four centuries have continued to evolve.

Telescopes are a very useful tool for studying distant objects. Telescopes gather and focus the light being emitted from these far off objects. Light encodes information about the object being observed. The technique used to study light being emitted, absorbed, or scattered when matter interacts with electromagnetic radiation is called spectroscopy. Astronomers can analyze the data and learn about its composition, temperature, and color. There are two types of telescopes; refracting and reflecting. Refracting telescopes work by bending light through a lens and forming an image, the kind that Galileo first used. Refraction happens when light passes through a glass prism. The lens collects the light by allowing all of the waves to pass through it focusing the waves on a spot at the other end of the tube. The image of the light emitting object is formed there and a smaller lens called the eyepiece at the back of the tube then magnifies the object. The problem with this type of telescope is that the design suffers from severe chromatic aberration. Therefore, a very large lens is needed to improve the magnification of this telescope.

The reflecting telescope was invented by Sir Issac Newton in 1668 as an alternative to refractors. Reflecting telescopes use a single or combination of curved mirrors to reflect the light and form an image. They have a large mirror (primary) at the back end. The curved primary reflects and focuses the light to a second smaller mirror (secondary) above it that gathers and redirects the light. There are different designs for this type of telescope such as a 49 Newtonian focus (secondary reflects light to the side of the telescope) or the Cassegrain focus (redirects the light back down to an opening in the center of the primary mirror). Reflecting telescopes are widely used in astronomy research. However, telescopes don't have to operate in just the visible (Optical) part of the electromagnetic spectrum; there are radio, infrared, UV, X-ray, and Gamma ray types. Due to earth's atmosphere preventing most of the incoming electromagnetic radiation, space-based telescopes eliminate problems of atmospheric turbulence and absorption. The benefit of having a telescope in space operating at these various wavelengths will be discussed in the next chapter.

GROUP ACTIVITY 3—DIFFRACTION OF LIGHT WAVES

Exercise 1

PROCEDURE 1

Open the PhET "Wave Interference"

Using the light tab, observe a single source of light waves.

Light waves are commonly denoted by their wavelength. Units: nanometers

1 nanometer = 1nm = 1×10^{-9} meter.

Use the slider on the color scale and the measuring tape to determine the following:

▶ Which color has the shortest wavelength?

Color_____   Wavelength_____

▶ Which color has the longest wavelength?

Color_____   Wavelength_____

▶ What changes do you observe when the amplitude increases?

PROCEDURE 2

Click on "Two Light Sources" and observe the Interference pattern.

Identify and describe locations of destructive interference.

How does the pattern change as the wavelength decreases?

PROCEDURE 3: MORE DETAILS ABOUT INTERFERENCE OF LIGHT WAVES

You may have noticed that the two light sources were "in sync"– technically termed "coherent". To replicate that arrangement with the PhET proceed as follow:

- ► *Slide the wavelength to blue*
- ► *Click on "one light" and "two slits"*
- ► *Set the barrier location midway between zero and 2590 nm*
- ► *Set the slit width one notch pass zero*
- ► *Set the slit separation midway between zero and 1750 nm*
- ► *Click on screen*
- ► *Click on measuring tape*

On the screen you should see a pattern of bright stripes and dark stripes. In the following you will want to observe how this pattern change due to changes in the wavelength of light, the slit width, the slit separation and the distance between the barrier and the screen. *Be sure to change only one variable at a time—return to the initial values before making the next change.*

In particular, you will use the measuring tape to measure the distance, x, from the center of one bright spot to the next. Initial value of $x =$ _____.

For each of the following indicate if x increases or decreases

- ► When the slit width increases, x *(increased, decreased, or stayed the same)*

- ► When the slit separation increases, x _____

- ► When the distance to the barrier increases, x _____

- ► When the wavelength of light increases, x _____

PROCEDURE 4: INTERFERENCE USING A LASER AND SLIT PLATE

Although the PhET simulations do a great job of modeling real phenomena, it might be both useful and fun to use real lasers and slits to perform the same experiment. (DO NOT LOOK DIRECTLY AT THE LASER LIGHT)

Use two lasers, one red and one green, and a slit plate with various slit widths and slit separations. It is important to note the initial value of the slid width, slit separation and the distance from the slit plate to the screen so that you can return to these values before making the next change.

Did your results agree with the PhET exercise? Yes _____ or No _____

Exercise 2

PROCEDURE 1

In this activity use both a PhET simulation and lab equipment (a laser, a slide with double slits, and a measuring tape) to investigate interference of light waves and thereby confirm that light behaves as a wave. Note the slide specifies separation between the slits, d.

Arrange the equipment so that an interference pattern is viewed on a distant wall or other vertical surface. Make 4 different measurements varying L.

Laser Color	d	L	Number of Bright Spots (1, 2, etc.)	X	Λ

Analysis and Questions

1. Compute the average of the 4 values you obtained for λ, and compare (evaluate % error; b-a/a x 100%) with the value provided by the manufacturer.

2. Do you think the value you obtained was reasonable given the circumstances (equipment and measuring techniques) of this experiment? Explain.

3. What changes in equipment or technique would you suggest for improving the results of this experiment? Be specific.

4. If blue light had been used, in what ways would you expect the interference patterns you observer differ from those of green and red light? Explain.

5. Why does this experiment demonstrate that light has wave properties?

6. What do you think the pattern on the wall might look like if light exhibited particle rather than wave properties?

1. What is light?

2. What is diffraction?

3. Where does light actually come from?

4. How can we prove that light bends?

5. What is the electromagnetic spectrum?

6. Could an experiment similar to Young's two-slit experiment be performed with sound? How might this be carried out? Does it matter that sound waves are longitudinal and electromagnetic waves are transverse? Explain.

7. Two coherent sources A and B of radio waves are 5.0 m apart. Each source emits waves with wavelength 6.0 m. Consider points along the line between the two sources. At what distances, if any, from A is the interference a) constructive and b) destructive?

8. Soapy water is colorless, but when blown into bubbles it shows vibrant colors. How does the thickness of the bubble walls determine the particular colors that appear?

9. You shine a tunable laser whose wavelength can be adjusted by turning a knob on a pair of closely spaced slits. The light emerging from the two slits produces an interference pattern on a screen like that shown in Figure 4.4. If you adjust the wavelength so that the laser light changes from red to blue, how will the spacing between bright fringes change?
   a. the spacing increases
   b. the spacing decreases
   c. the spacing is unchanged
   d. not enough information to make a decision

CHAPTER SUMMARY

In Group Activity 3 exercise 1 while performing the experiment with a monochromatic light source, you may have observed the light and dark bands created with the diffracting grating slide. Careful observation would have revealed the position of the geometric shadow line. The area outside the geometric shadow is bordered by alternating bright and dark bands. There is some light in the shadow region, although this is not very visible in Figure 4.4. The first bright band in Figure 4.4, just to the right of the geometric shadow, is considerably brighter than in the region of uniform illumination to the extreme right. This simple experiment gives us some idea of the richness and complexity of what might seem to be a simple idea, the casting of a shadow by an opaque object. We don't often observe diffraction patterns in everyday life because ordinary light sources are not monochromatic and are not point sources. If we use an incandescent light bulb instead of a monochromatic point source, each wavelength of the light from every point of the bulb forms its own diffraction pattern, but the patterns overlap to such an extent that we can't see any individual pattern.

You have also learned what happens with diffraction from a single slit; the pattern formed by plane-wave (parallel-ray), i.e., monochromatic light, when it emerges from a long, narrow slit (diffracting grating), the beam spreads out vertically after passing through the slit. The diffraction pattern consists of a central bright band, which may be much broader than the width of the slit, bordered by alternating dark and bright bands with rapidly decreasing intensity. Approximately 85% of the power in the transmitted beam is located in the central bright band, whose width is found to be inversely proportional to the width of the slit; the smaller the width of the slit, the broader the entire diffraction pattern.

We have analyzed interference from two point sources or from two very narrow slits; in this analysis we ignored effects due to the finite (that is, nonzero) slit width. We also considered the diffraction effects that occur when light passes through a single slit of finite width. Additional interesting effects occur when we have two slits with finite width or when there are several very narrow slits. Let's take another look at the two-slit pattern in the more realistic case in which the slits have finite width. If the slits are narrow in comparison to the wavelength, we can assume that light from each slit spreads out uniformly in all directions to the right of the slit. We used this assumption to calculate the interference patters as described in Exercise 1 by using equation 4.2, consisting of a series of equally spaced, equally intense maxima. However, when the slits have finite width, the peaks in the two-slit interference pattern are modulated by the single-slit diffraction pattern characteristic of the width of each slit. The effect of the finite width of the slits is to superimpose the two patterns—that is, to multiply the two intensities at each point. The two—slit peaks are in the same positions as before, but their intensities are modulated by the single-slit pattern, which acts as an "envelope".

The following terminology is used when discussing light waves:

▶ **Coherent**—two monochromatic sources of the same frequency and with any definite, constant phase relationship (not necessarily in phase).

▶ **Constructive Interference**—The crests or troughs of two interfering waves meet at the same point and are in phase, their amplitudes add together creating a single wave.

▶ **Destructive Interference**—When the crest of one wave meets the trough of the other. They are usually out of phase by ½ wavelength.

- ▶ **Diffraction Grating**—an array of a large number of parallel slits, all with the same width and spaced equal distances between centers.

- ▶ **Huygens's Principle**—Every point of a wave front can be considered the source of secondary wavelets that spread out in all directions with a speed equal to the speed of propagation of the wave. The position of the wave front at any later time is the envelope of the secondary waves at that time. To find the resultant displacement at any point, we combine all the individual displacements produced by these secondary waves, using he superposition principle and taking into account their amplitudes and relative phases.

- ▶ **Monochromatic light**—light of a single color, in optics they are characteristic of sinusoidal waves.

- ▶ **Polarization**—a characteristic of all transverse waves. The direction of polarization of a linearly polarized electromagnetic wave is the direction of the electric field. A polarizing filter passes waves that are linearly polarized along its polarizing axis and blocks waves polarized perpendicularly to that axis as in polarized sunglasses.

- ▶ **Principle of Superposition**—When two or more waves overlap, the resultant displacement at any point and at any instant is found by adding the instantaneous displacements that would be produced at the point by the individual waves if each were present alone.

CHAPTER 5

Geometric Optics

Your reflection in a mirror, the view of the moon through a telescope, and patterns seen in a kaleidoscope are all examples of images. In each of these cases the object that you are looking at appears to be in a different place than its actual position. For instance, your reflection in the mirror is on the other side of the mirror, the moon appears to be closer when seen through a telescope, and objects in the kaleidoscope seem to be in many places at the same time. In each case, light rays that come from a point on an object are deflected by reflection or refraction (or a combination of the two), so they converge towards or appear to diverge from a point called an image point. In Group Activity 4A—"bending Light", students will explore the laws of refraction; how much the refracted beam's direction differs from the incident beam depending on the medium.

Students will also get a sense of how different kinds of images form with simple optical devices in the PhET Sims "geometric optics", by using the laws of reflection, some simple geometry, and trigonometry are needed to understand images and image formation. The key role played by geometry in our analysis is the reason for the name *geometric optics* that is given to the study of how light rays form images. Before discussing what is meant by an image, the first thing needed is to understand what is the concept of an *object* as it is used in optics. By an *object* we mean anything from which light rays radiate. This could be emitted by the object itself; self-luminous, i.e., our sun or the glowing filament of a light bulb. The light could also be reflected sunlight, such as with the case of our moon. Light is a vital part of our everyday life. Take a look around you right now, what do you see? You could make a list of all the objects around you, but actually all you're really seeing is light that has interacted with those objects. Using your intuition and personal experiences, you are able to interpret color, patterns of light, and turn this into practical information. Knowing how to extract the maximum amount of information from light requires a deeper understanding of what light is and how it interacts with matter. How do we experience light? We've learned that light is a form of energy. Even with your eyes closed while you lie on the beach feeling the sunlight bathe your body and absorbs sunlight, you can feel the sunlight's intensity. If you remember from former science classes that greater warmth means more molecular motion, you'll realize that sunlight must be transferring its energy to the molecules in your skin. This is just one of the many ways in which we experience light. You can probably come up with a few other ways in which light plays an active role in day-to-day life. For example, daily experiences tell us that visible light is only part of the electromagnetic spectrum. More than likely you've seen a prism split light into its component colors as a rainbow of light called a spectrum as shown in Figure 5.1 below. In this example, the excited atoms within a hot dense object give off light of all colors (wavelengths) and produce a continuous spectrum—a complete rainbow of colors (range of wavelengths).

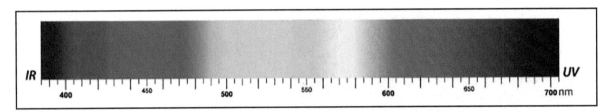

FIGURE 5.1   Continuous Spectrum

© Fouad A. Saad/Shutterstock.com

The basic colors in a rainbow, as shown in the continuous spectrum example are red, orange, yellow, green, blue, indigo, and violet. The common acronym learned in grade school ROYGBIV is familiar. We see white light when the colors are mixed in roughly equal proportions. Light from the sun or a light bulb is called white light because it contains all of the colors in the rainbow (The concept of a continuous spectrum will be discussed further in section 5.7 along with the entire electromagnetic spectrum). Black is what we perceive as the absence of light, therefore, no color. The wide variety of color comes from mixing just a few colors in varying proportions. Your television takes advantage of this fact to simulate a huge range of colors by combining only red, green, and blue light. These are known as the primary colors of vision, because they are the colors directly detected by the cells in your eyes. Colors tend to look different on paper, so artists generally work with an alternate set of primary colors; red, yellow, and blue. If you do any graphic design work, you may be familiar with the CMYK process that mixes the four colors cyan, magenta, yellow, and black to produce a great variety of colors. You can produce a spectrum either with a prism or a diffraction grating. We used diffraction gratings with a monochromatic laser to create constructive and destructive interference on a screen showing the pattern as a result. But did you know that a DVD or CD is also a diffraction grating? The bottom of a CD or DVD is etched with many closely spaced circles and therefore acts like a diffraction grating. That's why you see rainbows of color on the bottom of the disc when you hold it up to the light. The next time you have one in your hand, take a close look.

5.1   HOW DO LIGHT AND MATTER INTERACT?

When you look up at the night sky, the light coming from stars carry an enormous amount of information about the object's location, shape and structure, and composition to be decoded by your brain. You acquire this information when light enters your eyes, where special cells in your retina absorb it and send signals to your brain. Your brain interprets the messages that light carries, recognizing materials and objects in a process we call vision. All the information that light brings to Earth from the universe was encoded by one of the four basic interactions between light and matter. If you think about the interactions between light and matter that you see in everyday life, you will realize that light can interact through: 1) *emission*, 2) *absorption*, 3) *transmission*, and 4) *reflection/scattering*. Figure 5.2 focuses on the Law of Reflection, such as a plane mirror or lenses. In the case of a mirror, reflection reflects light along a simple path. The angle at which the light strikes the mirror is the same angle at which it is reflected. We will also discuss refraction later in the chapter. Refraction occurs when light travels through an area of space that has a changing index of refraction, such as the transition from air to water or light passing through a prism.

For now we will concentrate on how images are produced by *reflection*. We must reiterate what an object is by that we mean anything from which light rays radiate. In Figure 5.2, shown is the law of reflection and explains the effect of reflection from a plane mirror. Therefore, the reflection of the wave bounces off the surface, it reflects. The angle of reflection follows the formula: angle of incidence = angle of reflection. Where the *angle of incidence* is the angle between the incoming ray and the normal, and the reflected ray is the angle between the reflected ray and the normal. Since the surface is plane, the normal is perpendicular and in the same direction at all points on the surface. Many communications technologies use reflection of electromagnetic waves to facilitate the transfer of information such as CD-ROMs (Compact disc read-only memory) that can be

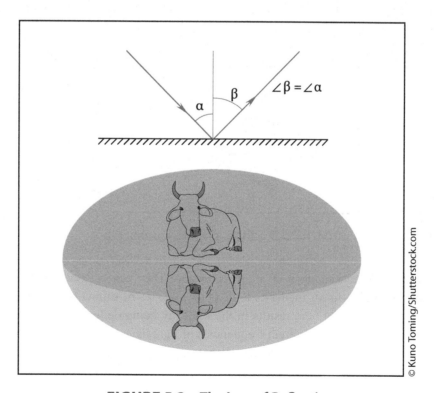

FIGURE 5.2   The Law of Reflection

read by a computer with an optical drive. If you want to store digital pictures on it, your computer must have a CD burner. There are many applications using the Law of Reflection, such as satellite dish in which case a parabolic reflector collects energy from a distant source such as light, sound or radio waves, security cameras where the mirrors are convex either round or rectangular in shape. They widen the field of view in areas hard to see around.

Refraction unlike reflection is related to the velocities of a wave in different media. We can outline how this may result in the bending of a wavefront. Refraction occurs when light travels through an area of space that has a changing index of refraction. The simplest case of refraction occurs when there is an interface between two uniform media with an index of refraction different than the other medium such as air vs. water. We can describe this situation using Snell's Law which states: the angles between the normal (to the interface) and the incident and refracted waves. We can write the formula, which describes the resulting deflection of the ray as:

$$n_1 \sin \theta_1 = n_2 \sin \theta_2 \hspace{3cm} \textbf{Equation 5.1}$$

where n_1 the index of refraction for uniform medium 1 (air) and n_2 the index of refraction for medium 2 (water), and θ_1 and θ_2 are the angles between the normal (to the interface) and the incident and refracted waves, respectively. One of the consequences of Snell's Law in everyday life is the use of the technology fiber optics. How does fiber optics work? If light leaving a slower medium at an incident angle greater than the critical angle, it won't refract, it won't be able to escape into the faster medium. Instead it will travel at the surface of the slower medium and will reflect at the boundary of the two media. Fiber optic cables work because the incident angles are so large they just keep reflecting within the fiber optic tube. Light rays traveling from a material with a high index of refraction to a material with a low index of refraction makes it possible for the interaction with the interface to result in zero transmission. In other words, as light travels down a fiber optic cable, it undergoes total internal reflection allowing for essentially no light to be lost over the length of the cable. Figure 5.3 is just one example of fiber optic cable. This particular picture is a fiber optic patch cable, Singlemode Duplex Fiber for long distance transmission.

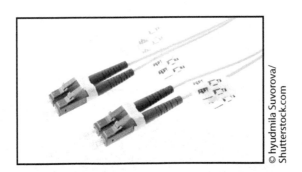

© hyudmila Suvorova/ Shutterstock.com

FIGURE 5.3 **Single Mode Duplex Fiber**

Prisms are an interesting consequence of Snell's Law too. They can be used to predict the deflection of light rays as they pass through a linear media as long as the indexes of refraction and geometry of the media are known. The propagation of light here results in the light ray being deflected depending on the shape and orientation of the prism. Additionally, different frequencies of light have slightly different indexes of refraction in most materials; refraction can be used to produce dispersion spectra as seen in Figure 5.1. The discovery of this phenomenon, passing light through a prism was attributed to Sir Isaac Newton.

5.2 PROPERTIES OF A LENS

The most familiar and widely used optical device after the plane mirror is the lens. A *lens* is an optical system with two refracting surfaces. The simplest lens has two spherical surfaces close enough together that we can neglect the distance between them (thickness). If you wear eyeglasses, as I do, or contacts lenses while reading, you are viewing these words through a pair of thin lenses. A lens of the shape as shown in Figure 5.4, has the property that when a beam of rays parallel to the axis passes through the lens, the rays converge to a point F on the right and forms a real image at that point. Such a lens is called a *converging lens*. Similarly, rays passing through point F on the left hand side of the lens emerge from the lens as a beam of parallel rays. The two points are called the first and second focal points, as you will discover when you play the geometric optics PhET Sims. The distance *f* is measured from the center of the lens to either of the focal points. In either case, the measured length is always the same for both sides of the lens. This measurement is called the *focal length*. The focal length of a converging lens is defined to be a *positive* quantity and therefore called a *positive lens*. The center horizontal line in Figure 5.4 is called the *optic axis*. The center of curvature of the lens' surface lies on and is defined as the optic axis. The two focal lengths, from the center of the lens (dashed vertical line) to either F's are always equal for a thin lens even when the two sides have different curvatures.

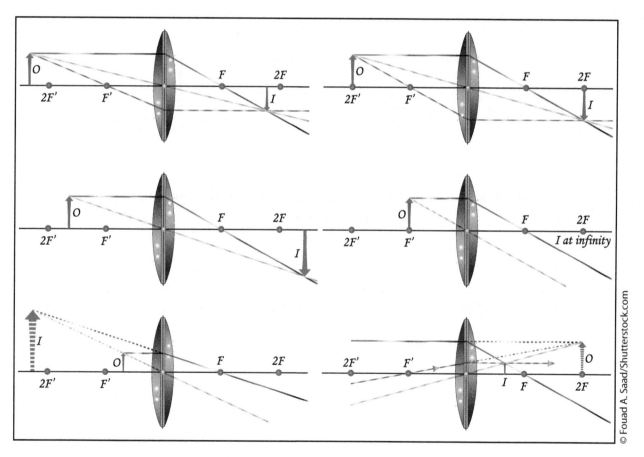

FIGURE 5.4   Thin Converging Lens

In Group Activity 4A, part B—Converging Lens, you will discover that a converging lens, like a concave mirror, can form an image of an extended object. Figure 5.5 shows how to find the position and lateral magnification of an image made by a thin converging lens. Before proceeding any further, let us introduce some general *sign rules*. These may seem unnecessarily complicated for the simple case of an image formed by a plane mirror, but we want to state the rules in a form that will be applicable to all the situations in Activity 4A. These will include image formation by a plane or spherical reflecting or refracting surface, or by a pair of refracting surfaces forming a lens. Here are the rules:

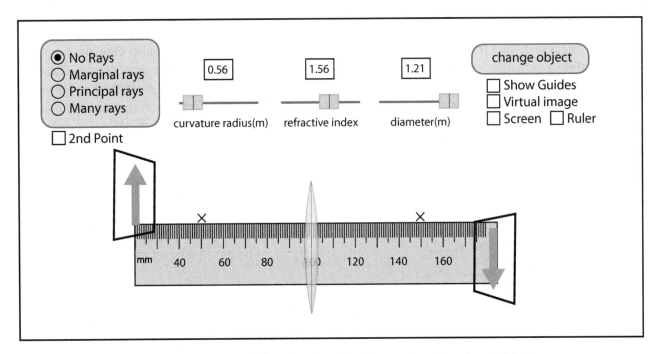

FIGURE 5.5   Construction Used to Find Image Position for a Thin Lens

© Kendall Hunt Publishing Company

1. **Sign rule for the object distance**: When the object is on the same side of the reflecting or refracting surface as the incoming light, the object distance is positive; otherwise, it is negative.
2. **Sign rule for the image distance**: When the image is on the same side of the reflecting or refracting surface as the outgoing light, the image distance is positive, otherwise it is negative.
3. **Sign rule for the radius of curvature of a spherical surface**: When the center of curvature is on the same side as the outgoing light, the radius of curvature is positive, otherwise, it is negative.

5.3 CAMERAS

The concept of image is so essential in understanding a simple mirror or lens system that it plays an equally important role in the analysis of optical instruments known as optical devices. One such optical device familiar to you is the camera. It makes an image of an object and records it either electronically or on film.

The basic elements of a camera are a light-tight box ("camera" is a Latin word meaning "a room or enclosure"), a converging lens, a shutter to open the lens for a prescribed length of time, and a light-sensitive recording medium, as shown in the diagram below (Figure 5.6). In a digital camera this is an electronic detector called a charged-couple device (CCD) array; in an older camera this is photographic film. The lens forms an inverted real image on the recording medium of the object being photographed. High quality camera lenses have several elements, permitting partial correction of various aberrations (something that deviates from the normal way), including the dependence of index of refraction on wavelength and the limitations imposed by the paraxial (a ray that makes a small angle to the optical axis of the system, and lies close to the axis throughout the system) approximation.

© NE Studio/Shutterstock.com

FIGURE 5.6   Schematic Diagram of a Digital Camera

When the camera is in proper focus, the position of the recording medium coincides with the position of the real image formed by the lens. The resulting photograph will then be as sharp as possible. With a converging lens, the image distance increases as the object distance decreases. Hence, in focusing the camera, we have to move the lens closer to the film for a distant object and farther from the film for a nearby object. In terms of focal length f, for a camera lens, it all depends on the film size and the desired angle of view. You can easily change the focal length of a camera. A lens with a long focal length is called a telephoto lens. This gives a small angle of view but a large image of a distant object. A lens with a short focal length gives a small image and a wide angle of view and is called a wide-angle lens. To understand this concept, recall that the focal length is the distance from the lens to the image when the object is infinitely far away. In general for any object distance, using a lens of longer focal length gives a greater image distance. This also increases the height of the image; the ratio of the image height to the object height which is equal in absolute value to the ratio of image distance to the object distance. With a lens of short focal length, the ratio image distance to the object distance is small, and a distant object gives only a small image. When a lens with a long focal length is used, the image of this same object may entirely cover the area of the film. Hence the longer the focal length, the more narrow the angle of view.

5.4 CAMERA LENSES AND THE *f*-NUMBER

For the film to record the image properly, the total light energy per unit area reaching the film (the exposure) must fall within certain limits. The shutter and the lens aperture control this. The shutter controls the time interval during which light enters the lens. This is usually adjustable in steps corresponding to factors of about 2, often from 1 second to 1 one-thousandth of a second. The intensity of light reaching the film is proportional to the area viewed by the camera lens and to the effective area of the lens. The size of the area that the lens "sees" is proportional to the square of the angle of view of the lens, and is proportional to roughly $1/f^2$. The effective area of the lens is controlled by means of an adjustable lens aperture, or *diaphragm*, a nearly circular hole with variable diameter D; hence the effective area is proportional to D^2. Putting these factors together, we see that the intensity of light reaching the film with a particular lens is proportional to D^2/f^2. The light-gathering capability of a lens is commonly expressed by photographers in terms of the ratio of f/D, called the *f*-number of the lens.

Many photographers use a variety of lenses, which have a variable focal length. The most popular is the *zoom lens*, which is not a single lens but a complex collection of several lens elements that give a continuously variable focal length, over a range as great as 10 to 1. Zoom lenses give a range of image size of a given object. It is an enormously complex problem in optical design to keep the image in focus and at the same time maintain a constant *f-number* whiles the focal length changes. When you vary the focal length of a typical zoom lens, two groups of elements move within the lens and a diaphragm opens and closes.

Other devices such as a projector for viewing slides, digital images, or motion pictures operate very much like a camera except in reverse. In a movie projector, a lamp shines on the film, which acts as an object for the projection lens. The lens forms a real, enlarged, inverted image of the film on the projection screen. Because the image is inverted, the film goes through the projector upside down so that the image on the screen appears right-side up.

5.5 THE HUMAN EYE

The optical behavior of the human eye is similar to that of a camera. The essential parts of the eye are show in Figure 5.7. It is considered as an optical system in itself. The eye is nearly spherical as the diagram shows and is about 2.5 cm in diameter. The front portion is somewhat more sharply curved and is covered by a tough, transparent membrane called the *cornea*. The region behind the cornea contains a fluid called the *aqueous humor*. Next in line is the *crystalline lens*, a capsule containing a fibrous jelly, hard at the center and progressively softer at the outer portions. The crystalline lens is held in place by ligaments that attach it to the *ciliary* muscle, which encircles it. Behind the lens, the eye is filled with a thin watery jelly called the *vitreous humor*. The indexes of refraction of both the aqueous humor and the vitreous humor are about 1.34, nearly equal to that of water (1.33). The crystalline lens, while not homogeneous, has an average index of refraction of 1.44. This is not very different from the indexes of the aqueous and vitreous humors. As a result, most of the refraction of light entering the eye occurs at the outer surfaces of the cornea.

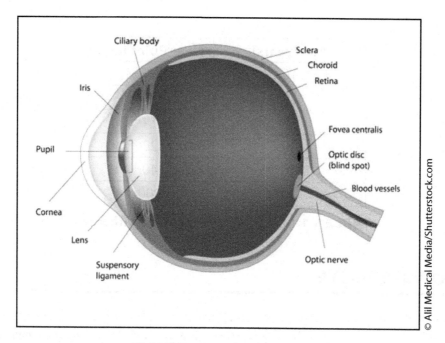

FIGURE 5.7   The Human Eye

Refraction at the cornea and the surfaces of the lens produces a *real image* of the object being viewed. This image is formed on the light-sensitive *retina*, lining the rear inner surface of the eye. The retina plays the same role as the film in a camera. The *rods* and *cones* in the retina act like an array of miniature photocells; they sense the image and transmit it via the *optic nerve* to the brain. Vision is most acute in a small central region called the *fovea centralis*, about 0.25 mm in diameter. In front of the lens is the *iris*. It contains an aperture with variable diameter called the pupil, which opens and closes to adapt to changing light intensity. The receptors of the retina also have intensity adaptation mechanisms.

For an object to be seen sharply, the image must be formed exactly at the location of the retina. The eye adjusts to different object distances by changing the focal length of its lens; the lens-to-retina distance does not change (Contrast this with focusing a camera, in which the focal length is fixed and the lens-to-film distance is changed). For the normal eye, an object at infinity is sharply focused when the ciliary muscle is relaxed. To permit sharp imaging on the retina of closer objects, the tension in the ciliary muscle surrounding the lens increases, the ciliary muscle contracts, the lens bulges, and the radii of curvature of its surfaces decrease; this decreases the focal length. This process is called *accommodation*.

The extremes of the range over which distance vision is possible are known as the *far point* and the *near point* of the eye. The far point of a normal eye is at infinity. The position of the near point depends on the amount of which the ciliary muscle can increase the curvature of the crystalline lens. The range of accommodation gradually diminishes with age because the crystalline lens grows throughout a person's life (it is about 50% larger at age 60 than at age 20) and the ciliary muscles are less able to distort a larger lens.

For this reason, the near point gradually recedes, as one grows older. This recession of the near point is called presbyopia. Table 5.1 shows the approximate position of the near point for an average person at various ages. For example, an average person 20 years of age cannot focus on an object that is closer than about 10 cm (test this for yourself).

Age (years)	Near Point (cm)
10	7
20	10
30	14
40	22
50	40
60	200

TABLE 5.1   Receding of Near Point with Age

5.6 VISION DEFECTS

Several common defects of vision result from incorrect distance relationships in the eye. A normal eye forms an image on the retina of an object at infinity when the eye is relaxed. In the *myopic* (nearsighted) eye, the eyeball is too long from front to back in comparison with the radius of curvature of the cornea (or the cornea is too sharply curved), and rays from an object at infinity are focused in front of the retina. The most distant object for which an image can be formed on the retina is then nearer than infinity. In the *hyperopic* (farsighted) eye, the eyeball is too short or the cornea is not curved enough, and the image of an infinitely distant object is behind the retina. The myopic eye produces too much convergence in a parallel bundle of rays for an image to be formed on the retina; the hyperopic eye, not enough convergence.

All these defects can be corrected by the use of corrective lenses; either eyeglasses or contact lenses. The near point of either a presbyopic or a hyperopic eye is farther from the eye than normal. To see clearly an object at normal reading distance (about 25cm), we need a lens that forms a virtual image of the object at or beyond the near point. This can be accomplished by using a converging (positive) lens, such as the one shown in Figure 5.4. In effect the lens moves the object farther away from the eye to a point where a sharp retinal image can be formed. Similarly, correcting the myopic eye involves using a diverging (negative) lens to move the image closer to the eye than the actual object.

Astigmatism is a different type of eye defect in which the surface of the cornea is not spherical any longer but rather more sharply curved in one plane than in the other. As a result, horizontal lines may be imaged in a different plane from vertical lines. Astigmatism may make it impossible, for example, to focus clearly on the horizontal and vertical bars of a window at the same time. Astigmatism can be corrected by the use of a lens with a cylindrical surface. For example, suppose the curvature of the cornea in a horizontal plane is corrected to focus rays from infinity on the retina but the curvature in the vertical plane is too great to form a sharp retinal image. When a cylindrical lens with its horizontal axis is placed before the eye, the rays in a horizontal plane are unaffected, but the additional divergence of the rays in a vertical plane causes these to be sharply imaged on the retina.

Lenses for vision correction are usually described in terms of the **power**, defined as the reciprocal of the focal length expressed in meters. The unit of power is the *diopter*. Thus, a lens with $f = 0.50$ m has a power of 2.0 diopters, a lens with $f = -0.25$ m corresponds to -4.0 diopters, and so on. The numbers on a prescription for glasses are usually powers expressed in diopters. When the correction involves both astigmatism and myopia or hyperopia, there are three numbers: one for the spherical power, one for the cylindrical power, and an angle to describe the orientation of the cylinder axis. An alternative approach for correcting many defects of vision is to reshape the cornea. This is often done using a procedure called *laser-assisted in situ keratomileusis*, or LASIK. An incision is made into the cornea and a flap of outer corneal tissue is folded back. A pulsed ultraviolet laser with a beam only 50 microns wide (about 1/200 the width of a human hair) is then used to vaporize away the microscopic area of the underlying tissue. The flap is then folded back into position, where it conforms to the new shape "carved" by the laser.

5.7   THE ELECTROMAGNETIC SPECTRUM REVEALED

At the beginning of this chapter the concept of a continuous spectrum was introduced. However, this is only a fraction of the entire spectrum. Yes, we see in the visible part of the spectrum but there is so much more information that can be obtained by studying the various wavelengths. Astronomers analyze the light that comes from far off objects. By decoding this information we can learn about the properties of these objects, especially by analyzing starlight. Here's a good analogy to how important it is for astronomers to analyze starlight, "it is as essential as you taking your next breath"! We've already learned that they can obtain a star's temperature and chemical composition. By looking at the peak wavelength of a star's blackbody curve, the temperature and color can be obtained as seen in Figure 5.8. Therefore, the excited atoms within a hot dense object will give off light of all colors (wavelengths), a complete rainbow of color across the entire range of wavelengths without any spectral lines. A Fluorescent tube light is a mercury-vapor discharge lamp. Fluorescence produces visible light which is found in many office environments.

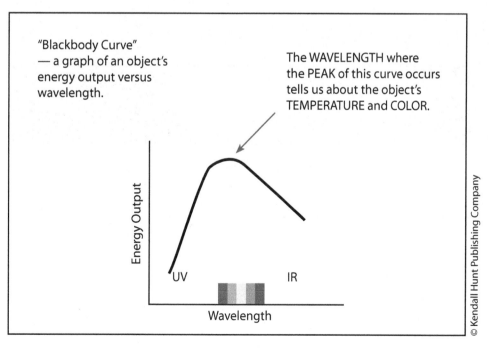

FIGURE 5.8   Wavelength vs. Energy Output of a Blackbody Curve

All objects; planets, stars, etc. have radiative energies that become randomized so that the energies are spread over a wide range of wavelengths. The wide range of photons explains why a spectrum looks smooth and continuous. Most spectrum of these objects depend on only one thing, the temperature. Photon energies depend only on temperature regardless of the composition! The temperature dependence of this light explains why we call it thermal radiation (blackbody radiation). Thermal radiation spectra are the most common type of continuous spectra.

Electromagnetic radiation comes from the acceleration of charged particles. The concept of light was debated in the 17th Century by both Isaac Newton (composed of little particles) and Christian Huygens (waves). However, in the 19th and 20th Century Maxwell, Young, Einstein and others were able to show that Light behaves both like a particle and a wave depending on how you observe it. As mentioned earlier, when all of the flavors (light) are put together, the result is white light. This is why we see the sun as white because all of the colors in the visible part of the spectrum under the blackbody curve are in equal proportion. The range of wavelengths in the visible spectrum are shown below in Figure 5.9. Our sun is just an ordinary star. The peak wavelength is 500 nanometers which corresponds to the color green. Like all other stars it produces absorption spectra. This happens when the light from a hot dense object passes through a cool cloud of gas, i.e. the photosphere and the atoms within the cloud absorb particular colors (wavelengths) of light. What is observed is a series of dark spectral lines on a rainbow background as shown in Figure 5.10.

FIGURE 5.9   Wavelengths of the Visible Spectrum

© Fouad A. Saad/Shutterstock.com

FIGURE 5.10   Absorption Spectra

Courtesy of NASA.gov

The spectra of our sun and most stars show numerous dark lines against a rainbow background. Why aren't they bright lines as in emission spectra? Because the Sun is intensely hot! The Sun's visible surface or **photosphere** is in fact too hot for its hydrogen atoms to retain their electrons in any orbit and the atoms become **ionized.** The energy of these atoms absorb the radiation produced deep within the Sun is released as photons of many wavelengths. The diagram below, Figure 5.11 is an example of the element hydrogen and its spectral lines. Elements such as hydrogen in the atmosphere absorb radiation only at wavelengths that allow their electrons to jump to higher orbits as seen in Figure 5.11 above according to the Bohr Model of the atom. This selective diminution of light at specific wavelengths produces an absorption spectrum. These lines are characteristic of the composition of the intervening gas which occurs at precisely the same wavelengths as the emission lines produced by the gas at higher temperatures. In some of the lab activities that are done in this course, students will have an opportunity to observe absorption spectra just as if you were analyzing real starlight.

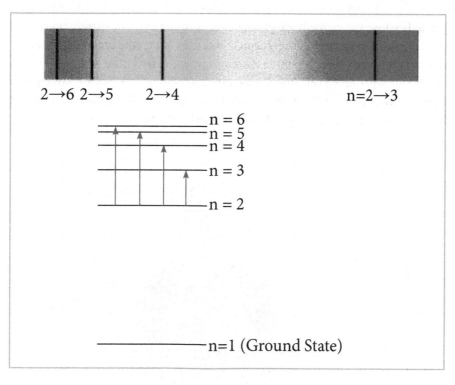

FIGURE 5.11   Spectral Lines of Hydrogen

Courtesy Gerceida Jones

A great light source to use is a neodymium light bulb to simulate the sun. The last to be discussed before moving on to look at the entire electromagnetic spectrum, is an emission spectrum. In this case, the excited atoms within a hot, cloud of gas give off only particular colors (wavelengths) of light. What is observed is a series of bright spectral lines against a dark background. Gas discharge tubes of various elements can be used to demonstrate how these will emit spectral lines that are unique only to their atoms. Again, each chemical element produces its own unique set of spectral lines when it is excited as show in Figure 5.12. In many High school chemistry classes, the flame test experiment with various salts will show the color of a particular

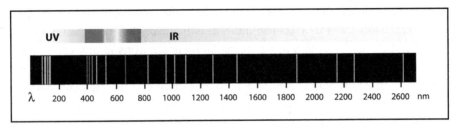

FIGURE 5.12 Emission Spectra of Hydrogen
© magnetix/Shutterstock.com

element when sprinkled into the flame as demonstrated in Figure 5.13. However, to see the spectral lines, a spectroscope will reveal the true nature of the element. This is how astronomers know what they are seeing in far off objects in space because the element looks the same as in the lab. The different energy levels of different atoms and molecules allow light to carry "fingerprints" that can tell us the chemical composition of those distant objects. Atoms contain (store) energy in three different ways: 1) by virtue of their mass (mass-energy)—E= mc², 2) Kinetic Energy (motion), and 3) electrical potential energy (encoded information) which depends on the arrangement of their electrons around the nuclei. In order to interpret the messages carried by light, we must understand how the electrons store and release their electrical potential energy.

FIGURE 5.13 The Flame Test

Matter leaves its fingerprints whenever it interacts with light. Light can interact with matter in four basic ways: 1) emission, 2) absorption, 3) transmission, and 4) reflection/scattering. Newton proved that the color came from the light by placing a second prism in front of the light of just one color. This tells us about the nature of color but not what light is. Young proved that light is both a particle (thing) and a wave (pattern) in his double-slit experiment and we know that carries energy. But did you know that scientists are able to use the entire electromagnetic spectrum to gather information from these energies of far off objects? Let's look at the spectrum in its totality and see how it is used to both benefit and harm mankind in Table 5.2.

TABLE 5.2 The Entire Electromagnetic Spectrum Revealed

All images © Shutterstock.com

Type of Wave	Wavelength	Benefits	Harmful Effects
Radio	200 to 600 m	Radio astronomy, listening to music, and talk radio	None that is known.
Microwaves	1 mm to 1 m	Microwave astronomy, food, cell phones, radar	At very high intensities heat up and kill living cells.
Thermal Infrared	10μ $1\mu = 10^{-6}$m	Restaurants, hospitals, animal vision, security systems	At very high intensity can heat up living tissue and kill them.
Near Infrared	0.8 to 3μ	Looking @ young stars, communications in both the air & fiber optics	This wavelength is generally safe but at very high intensity can heat tissue.
Visible	0.6μ to 0.4μ or 380nm to 740nm	Vision, Plant growth, Lasers	Normally harmless, can cause blindness or burn tissue at high intensity.
Ultraviolet A	0.34 to 0.4μ	Attracting insects, Illuminating black light posters, and Mineral Identification	In high doses may contribute to skin cancer or eye damage.

TABLE 5.2 *Continued.*

Type of Wave	Wavelength	Benefits	Harmful Effects
Ultraviolet B	0.29 to 0.32µ or 290nm to 320nm	Studying the sun & hotter stars and tanning beds	Sunburn and skin cancer
X-rays	10^{-10} m < 1 nm the length of a water molecule to the length of a large protein molecule	Astronomical observations, medical diagnosis, security scanning.	DNA mutations, high doses can cause death, lower doses can cause cancer.
Gamma Rays	0.001nm and shorter; much smaller than atom	Studying gamma-ray bursts, detecting radioactivity, medical treatments Compton Gamma Ray Astronomical Observatory satellite Detecting Nuclear Weapon Explosions, such as those tested by North Korea!	Cancer, radiation sickness

GROUP ACTIVITY 4A

Procedure I

REFRACTION

Open the PhET (http://phet.colorado.edu) and the simulation "Bending Light." On the Intro tab click on the Laser View—Ray and start the laser.

Adjust the Materials so that the upper is Air and the lower is Water.

Look at the *refracted* beam, materials vary in *index of refraction*—which determines how much the refracted beam's direction differs from the incident beam.

Vary the materials and observe the relationship between the incident and refracted rays for various combinations of upper and lower indices of refraction.

In your observations, which combinations of indexes of refraction produce the most "bending" and which produce the least?

Procedure II

CONVERGING LENS

Open the PhET simulation "Geometric Optics."

Set the simulation for "no rays" and make sure that the "screen" is *not* checked.

Definition: The distance between the center of the lens and the focal point is called the *focal length* of the lens.

Play with the distance between the object and the lens, what do you observe?

As you vary the curvature and refractive index of the lens and describe the changes you see in *focal length*.

How would you construct a lens to have the shortest or longest focal length?

Now click on "marginal rays—vary the distance between the object and lens, describe the relationship between object distance and image.

Vary the curvature. Describe qualitatively the relationship between the curvature and the image distance.

Procedure III

CAMERAS AND EYES

This arrangement of object, lens and screen in this simulation describes the basic operation of a camera or an eye. Use the PhET simulation to help answer the questions below.

Describe how a camera is able to focus on objects that are near and far.

Compare how a camera focuses vs. the human eye.

Procedure IV

REAL LENSES

Obtain a "real" glass lens, preferably 100mm or 200mm and measure its focal length by choosing a very distant object that you bring into focus on a screen as demonstrated by Wendy in Figure 5.14. Describe whether the image is upright or inverted, measure the focal length, and then compare your results to manufacturer's value. What is the percent error? Why did your value vary differently from the manufacturer's value?

Observe the curvature of the two lenses. What is the relationship you observe between curvature and focal length and how does that compare with the PhET observations in procedure II?

Use a lens and produce an image on the screen for a nearby object making sure that the object is beyond the focal point of the lens. Describe qualitatively the relationship between object distance and image distance for the lens. Does this little experiment agree with the PhET observations in the simulation?

Using two different lens, compare the image distance for a given object distance. Does this agree with the PhET simulation?

Courtesy of Gerceida Jones

FIGURE 5.14   Wendy compares the PHeT simulation results to real lenses.

GROUP ACTIVITY 4B

Procedure I

SPECTROSCOPE

Look through the spectroscope at an incandescent light bulb and observe the colors of the rainbow. Record the wavelengths.

Procedure II

FILTERS

Hold a red filter in front of the spectroscope as demonstrated by Nathaniel and Donna in Figure 5.15. Filters take out or absorb particular wavelengths. Look to your right, parts of the spectrum are missing or dim. Now try other filters, what do you see? Make a table of filter color, colors that are transmitted (pass through the filter), and colors absorbed (those that don't). Check the wavelengths and compare to what you observed in procedure I.

Courtesy of Gerceida Jones

FIGURE 5.15   Donna and Nathaniel use various colored filter for Procedure II.

Procedure III

PRIMARY COLORS OF LIGHT

The spectroscope is used to determine the wavelengths of the different colors that make up white light. Determine the wavelengths of the primary colors.

The red, blue and green filters should be used because adding these colors in varying amounts will produce all the colors of light. They are also good approximations of **ideal primary filters**. An **ideal primary filters** can transmit 100% at some wavelengths and 0% at all other wavelengths.

The primary colors of light are called *additive primaries* and are not the same as primary colors of paint which are *subtractive primaries*.

The *subtractive primaries* are magenta, cyan and yellow.

Additive mixing of light happens when two or more colors of light are superimposed on top of each other. As more and more light is superimposed, what results is brighter light and our brain perceives a color that is different from the colors of the original beam as you will discover in the PhET "Color and Vision".

Procedure IV

Open the PhET "Color Vision". Using the RGB tab, add equal amounts of Red, Green and/or Blue and observe the resulting color. Make a table colors added vs. the resulting color.

By playing with the PhET simulation answer the following questions:

► What color is the result of adding primary green to the magenta?

► What are complementary colors?

► What color is observed when red light is overlapped with green light? What color is complementary to yellow?

► What color is observed when blue light is overlapped with green light? What color is complementary to cyan?

Use the diagram below to indicate the colors observed by the overlapping portions shown.

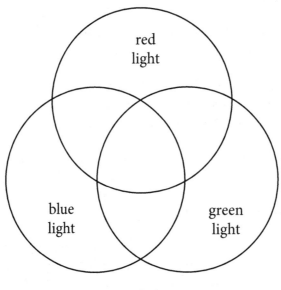

What is partitive mixing of light?

What is subtractive mixing of light?

Color printing of magazines, books, etc. is a result of both additive and subtractive color mixing. Take a close look at newspaper cartoons with a magnifying glass and record what you observe in subtractive and additive mixing.

How do black and white prints differ from color printing? Observe black and white newspaper cartoons with a magnifying glass and record your observations. Is there a difference between the black areas versus the gray areas?

QUESTIONS AND ANSWERS

1. Explain why the focal length of a plane mirror is infinite, and explain what it means for the focal point to be at infinity.

2. If a spherical mirror is immersed in water, does its focal length change? Explain.

3. A person looks at his reflection in the concave side of a shiny spoon. Is it right side up or inverted? Does it matter how far his face is from the spoon? What if he looks in the convex side? (Try this yourself!)

4. The bottom of the passenger side mirror on your car notes, "Objects in mirror are closer than they appear". Is this true? Why?

5. How could you very quickly make an approximate measurement of the focal length of a converging lens? Could the same method be applied if you wished to use a diverging lens? Explain.

6. When a converging lens is immersed in water, does its focal length increase or decrease in comparison with the value of air? Explain.

7. A candle 4.85 cm tall is 39.2 cm to the left of a plane mirror. Where is the image formed by the mirror, and what is the height of this image?

8. A speck of dirt is embedded 3.50 cm below the surface of a sheet of ice (n_1 = 1.309). What is its apparent depth when viewed at normal incidence?

9. Which has the greatest potential energy, electrons in inner electron orbits or electrons in outer orbits?

10. In your laboratory observations, which of the photons that you observed had the highest energy?

11. How can an electron be made to jump to a higher energy level?

12. What happens when an "excited" electron falls down to a lower energy level?

13. It is often stated that a spectrum is a "chemical fingerprint." What do you think this means?

14. Imagine an emission spectrum produced by a container of hydrogen gas. One of your classmates argues that changing the amount of hydrogen in the container will change the colors of the lines in the spectrum. Is he right? Explain.

15. How do you think scientists might determine what elements are in distant stars?

16. Consider the following diagram of a hypothetical atom. An electron normally orbiting in orbit #2 is "kicked" into orbit #5. Describe the different ways in which it can return to orbit #2. In which way will it emit the higher energy photon?

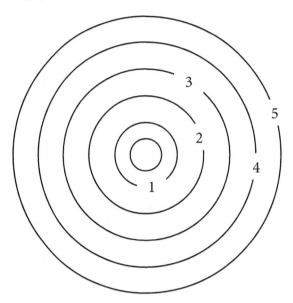

© Kendall Hunt Publishing Company

17. Pick three of the different ways that the electrons can return to orbit #2 in this hypothetical atom. If these resulted in the emission of photons that correspond to red, blue and green, which of the ways would correspond to each of these colors? Indicate the electron transition (e.g. 3 → 2) and the color.

18. Imagine that you observe the Sun using a telescope in an orbit high above Earth's atmosphere. Which type of spectra would you observe by analyzing sunlight?
   a.   Dark line absorption Spectrum
   b.   Bright line emission spectrum
   c.   Continuous spectrum
   d.   None of the above

19. If an electron in an atom moves from an orbit with an energy of 5 to an orbit with an energy of 10
   a.   A photon of energy 5 is emitted
   b.   A photon of energy 15 is emitted
   c.   A photon of energy 5 is absorbed
   d.   A photon of energy 15 is absorbed

CHAPTER SUMMARY

The following terminology is used when discussing optics:

- **Absorption**—matter can capture electromagnetic radiation and convert the energy of a photon into internal energy

- **Angle of Incidence**—the angle between the incoming ray and the normal, and the reflected ray is the angle between the reflected ray and the normal

- **Converging lens**—a lens (curved piece of glass) that bends light so that it can produce an image

- **Emission**—atoms, molecules, or solids that are excited to high energy levels can decay to lower energy levels by emitting radiation

- **Focal point**—the point at which the incident parallel rays converge

- **Focal length**—an optical system is a measure of how strongly the system converges or diverges light. It is the distance from the vertex to the focal point, denoted by f and related to the radius of curvature.

- **Lenses**—Concave (thinner at the center and a parallel beam of light passing through the center will cause the light to spread out or diverge) versus Convex (thicker at the center and the light rays passing through will be brought closer together or converge)

- **Object**—anything from which light rays radiate

- **Real image**—the resulting image when outgoing rays pass through an image point

- **Reflection**—occurs when light changes direction as a result of "bouncing off" a surface like a mirror

- **Refraction**—the bending of light as it passes from one medium to another

- **Transmission**—the process in which light passes through a medium without being absorbed or scattered

- **Spectrometer**—measures the wavelengths of the light and how intense (powerful/bright) the light is at each wavelength.

© Oxana Gracheva/Shutterstock.com

CHAPTER 6

MAGNETIC FIELDS AND MAGNETIC FORCES

Everyone uses magnetic forces whether they are aware of it or not. Magnetic forces are at the heart of electric motors, TV picture tubes, microwave ovens, loudspeakers, computer printers, and disk drives. The most familiar aspects of magnetism are those associated with permanent magnets, which attract nonmagnetic iron objects and can also attract or repel other magnets. A compass needle aligning itself with the earth's magnetic field is an example of this interaction. But the fundamental nature of magnetism is the interaction of moving electric charges. Unlike electric forces, which act on electric charges whether they are moving or standing still, magnetic forces act only on moving charges. A magnetic field is also an electrostatic field moving at a very high or relativistic speed depending on the frame of reference.

Magnetic and electric forces are very different from one another. We use the concept of a field to describe both kinds of forces. Electric forces arise in two stages: 1) a charge produces an electric field in the space around it, and 2) a second charge responds to this field. Magnetic forces also arise in two stages; the first stage is a moving charge or a collection of moving charges (an electric current) that produces a magnetic field. The second stage is a current or moving charge that responds to this magnetic field, and therefore experiences a magnetic force.

In this chapter students will study the second stage in the magnetic interaction—that is, how moving charges and currents respond to magnetic fields. In particular, they will discover why magnets can pick up iron objects like paperclips and eventually learn and examine how moving charges and currents produce magnetic fields.

6.1   THE MAGNETIC EARTH

Magnetic phenomena were first observed some 2500 years ago in fragments of magnetized iron ore or load-stones found near the Greek province of Magnesia, now Manisa, Turkey. These fragments were examples of what are now called *permanent magnets*; you probably have several permanent magnets on your refrigerator door in your dorm, apartment, or home. Permanent magnets were found to exert forces on each other as well as on pieces of iron that were not magnetized. It was discovered that when an iron rod is brought in contact with a natural magnet, the rod also became magnetized. When such a rod is floated on water or suspended by a string from its center, it tends to line up in a north-south direction. The needle of an ordinary compass is just such a piece of magnetized iron.

Before the relationship of magnetic interactions to moving charges was understood, the interactions of permanent magnets and compass needles were described in terms of *magnetic poles*. If a bar-shaped permanent magnet, or *bar magnet*, is free to rotate, one end points north. This end is called a *North Pole* or *N pole*; the other end is a *South Pole* or *S pole*. There are no individual magnetic poles, magnetic charges or monopoles found in nature. Electric charges can be separated, but magnetic poles always come in pairs known as a dipole as seen in Figure 6.1. Opposite poles attract each other (Figure 6.1a & b), and like poles repel each other

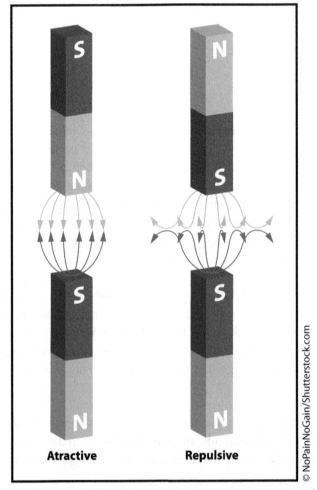

FIGURE 6.1   Bar Magnets (permanent magnets)

(Figure 6.1c &d). These bar magnets will remain permanent unless something happens to eliminate the alignment in the atomic magnets in the bar of cobalt (Co), iron (Fe), or nickel (Ni). An object that contains iron but is not itself magnetized (no tendency to point either north or south) is attracted by either pole of a permanent magnet. This is an example of the attraction that acts between a magnet and the nonmagnetic nail. By analogy to electric interactions, we can describe the interactions shown in Figure 6.1 a-d as a bar magnet that sets up a magnetic field in the space around it and a second body responds to that field. A compass needle tends to align with the magnetic field at the needle's position.

The earth itself is a magnet. Its north geographic pole is close to a magnetic south pole, which is why the north pole of a compass needle points north. The earth's magnetic axis is not quite parallel to its geographic axis (the axis of rotation), so a compass reading deviates somewhat from geographic north. This deviation, which varies with location, is called magnetic declination or magnetic variation. Also, the magnetic field is not horizontal at most points on the earth's surface; its angle up or down is called magnetic inclination. At the magnetic pole the magnetic field is vertical. In other words, earth's magnetic field appears to come from a giant bar magnet within but with its south pole located up near the earth's North geographic pole as shown in Figure 6.2 below. As you can see the magnetic field lines come out of earth near South America in Figure 6.2 and re-enter near Canada. These magnetic field lines show the direction that a compass would point at each location. The direction of the field at any point can be defined as the direction of the force that the field would exert on a magnetic north pole. The field, which is caused by the motion of currents in earth's molten outer core and a rapid rotation, changes with time. We have geologic evidence on the bottom of the ocean floor that the magnetic field reverses direction entirely at irregular intervals of about a half million years.

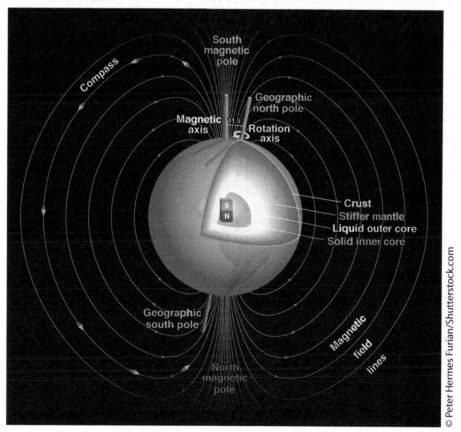

FIGURE 6.2 **Earth's Magnetic Field**

6.2 MAGNETIC POLES AND ELECTRIC CHARGES

The connection between an electric current and magnetic field was first observed when the presence of a current in a wire near a magnetic compass affected the direction of the compass needle. We know that currents give rise to magnetic fields, just as an electric charge will give rise to an electric field. The concept of magnetic poles with a north and south end may appear similar to that of an electric charge, with positive and negative charges, but the analogy can be very misleading. While isolated positive and negative charges do exist, there is no experimental evidence that a single isolated magnetic pole (mono-pole) exists; poles always come in pairs (dipole). If a bar magnet is broken into two pieces, each broken piece becomes a pole. The existence of an isolated magnetic pole, or magnetic monopole, would have sweeping implications for theoretical physics. Extensive searches for magnetic monopoles have been carried out, but none found as of yet.

The Danish scientist Hans Christian Oersted discovered the first evidence of the relationship of magnetism to moving charges in 1820. He found that a current-carrying wire as shown by Figure 6.3 deflected a compass needle. Andre Ampere carried out similar investigations in France. A few years later, Michael Faraday in England and Joseph Henry in the United States discovered that moving a magnet near a conducting loop can cause a current in the loop. We now know that the magnetic forces between two bodies as shown in Figure 6.1 are fundamentally due to the interactions between moving electrons (the spin) in the atoms of the bodies. There are also electric interactions between the two bodies, but these are far weaker than the magnetic interactions because the two bodies that are electrically neutral. Inside a magnetized body such as a permanent magnet, there is a coordinated motion of the atomic electrons but in a nonmagnetic body these motions are not coordinated. In other words, the electrons rotation and spin are random; they will cancel each other out.

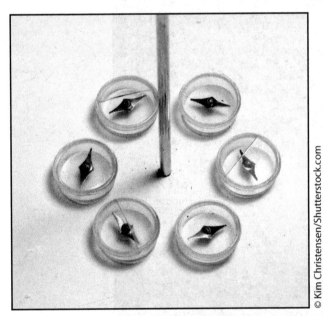
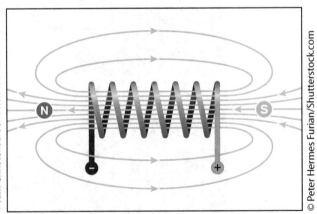

FIGURE 6.3   A compass near a current-carrying wire

Electric and magnetic interactions prove to be intimately connected as you have probably gathered by now. For the remainder of this book, we will develop the unifying principles of electromagnetism. Maxwell's equations would represent the synthesis of electromagnetism just as Newton's laws of motion are the synthesis of mechanics. However, we will cover as much as time allows.

6.3 MAGNETIC FORCES ON A MOVING CHARGE

There are four key characteristics of the magnetic force on a moving charge: 1) its magnitude is proportional to the magnitude of the charge. In other words, if the force on a charge q_1 and q_2 is doubled, as the charges move through a given magnetic field with the same velocity, experiments show that the force on q_2 is twice as great as the force on the q_1 charge. 2) The magnitude of the force is also proportional to the magnitude, or "strength," of the field; if we double the magnitude of the field, for example, by using two identical bar magnets instead of one without changing the charge or its velocity, the force doubles similar to what happens in Newton's Law by doubling the masses. 3) The magnetic force depends on the particle's velocity. This is quite different from the electric-field force, which is the same whether the charge is moving or not. A charged particle at rest experiences no magnetic force. And finally, 4) we find by experiment that the magnetic force vector **F** *does not* have the same direction as magnetic field vector **B** but instead is always *perpendicular to* both **B** and the velocity vector, *v*. Figure 6.4a–c on the next page shows these relationships. In Figure 6.4a, the magnitude F of the force on a charge is found proportional to when vector **v** and **B** are parallel, the force is zero. In other words, a charge moving parallel to a magnetic field experiences zero magnetic force. In Figure 6.4 b, the direction of F is always perpendicular to the plane containing *v* and B. Its magnitude is given by the equation:

$$F = |q|v \times B = |q|vB\sin(\varphi) \qquad \textbf{Equation 6.1}$$

Where $|q|$ is the magnitude of the charge (+ or −, a scalar quantity) and (phi, φ) is the angle measured from the direction of **v** to the direction of **B**, and where $v \times$ **B,** *v* is perpendicular to **B** is the cross product as shown in Figure 6.4. In other words, the cross product in Equation 6.1 represents the result that is perpendicular to both of those vectors and you will obtain a number. As mentioned above, if they are parallel, then the magnitude of the field will have no impact on the charge. In order to have an effect on the charge, the force has to be perpendicular to both the velocity of the charge and the magnetic field.

This description does not specify the direction of F completely; there are always two directions, opposite to each other, that are both perpendicular to the plane of *v* and **B**. To figure out the direction, we use the right-hand rule. The direction of the cross product can be explained by this method. Using the fingers of the right hand, point in the direction of the first vector *v*, in the cross product, then adjust your wrist so you can bend your fingers at the knuckles toward the direction of the second vector **B** and extend the thumb to get the direction of force. In other words, look at Figure 6.4b; imagine turning v until it points in the direction of B, turning through φ. Now wrap the fingers of your right hand around the line perpendicular to the plane of *v* and **B** so that they curl around with the sense of rotation from *v* to **B**. Your thumb then points in the direction of the force **F** on a positive charge. Alternatively, the direction of the force **F** on a positive charge is the direction in which a right hand thread screw as seen in Figure 6.4 (b) would advance if turned the same way.

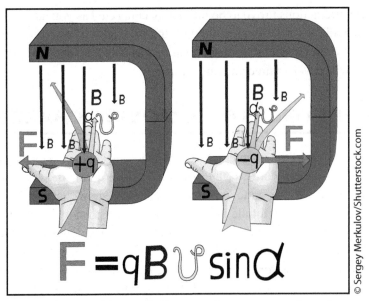

$$F = qB\upsilon \sin\alpha$$

FIGURE 6.4   Magnetic force acting on a moving charge

This discussion shows that the force on a charge q moving with velocity **v** in a magnetic field **B** is given, both in magnitude and in direction, by the following equation:

$$\mathbf{F} = q\mathbf{v} \times \mathbf{B} \qquad\qquad \textbf{Equation 6.2}$$

This equation is important because it is an observation based on experiment and not deduced theoretically. It is valid for both positive and negative charges. When q is negative, the direction of the force F is opposite to that of v × B. In other words, if two charges with equal magnitude and opposite sign move in the same B field with the same velocity, the forces have equal magnitude and opposite direction.

Equation 6.1 gives the magnitude of the magnetic force F in equation 6.2. We can express this magnitude in a different but equivalent way. Since we have the angle φ between the directions of vectors v and B, we may interpret **B**sinφ as the component of **B** perpendicular to v—that is, **B** perpendicular (**B**⊥). With this notation the force magnitude can be written as:

$$F = |q|v\mathbf{B}\perp \qquad\qquad \textbf{Equation 6.3}$$

This form is sometimes more convenient, especially in problems involving currents rather than individual particles. We will discuss forces on *currents* very shortly.

From Equation 6.1, the units of B must be the same as the units of F/qv. Therefore, the SI units of **B** is equivalent to 1 Newton–Second/Coulomb–meter, or, since one ampere is one coulomb per second (1 Ampere = 1 Coulomb/sec), 1 N/A–m where N (Newton), A (Ampere), and C (Coulomb). This unit is called the **tesla** (abbreviated **T, 1T = 10,000 Gauss**), in honor of Nikola Tesla (1857–1943), the prominent Serbian-American scientist and inventor:

$$1 \text{ tesla} = 1 \text{ T} = 1 \text{ N/A} - \text{m}$$

A **gauss** ($1G = 10^{-4}$ T) is also in common use. Instruments used for measuring magnetic field are sometimes called *gaussmeters*.

The magnetic field of the earth is on the order of 10^{-4} T or 1 G. Magnetic fields of the order of 10 T occur in the interior of atoms and are important in the analysis of atomic spectra discussed in chapter 5. The largest steady magnetic field that can be produced at present in the laboratory is about 45 T. Some pulsed-current electromagnets can produce fields of the order of 120 T for short time intervals on the order of a millisecond. The magnetic field at the surface of a neutron star is believed to be of the order of 10^8 T.

6.4 MEASURING MAGNETIC FIELDS

To explore an unknown magnetic field, we can measure the magnitude and direction of the force on a moving test charge and then use equation 6.2 to determine B. The electron beam in a cathode-ray tube, such as that used in a television set, is a convenient device for making such measurements. The electron gun shoots out a narrow beam of electrons at a known speed as seen in Figure 6.5. If there is no force to deflect the beam, it strikes the center of the screen.

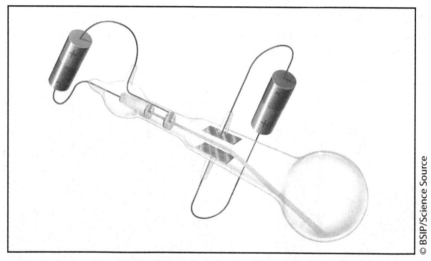

FIGURE 6.5 Cathode Ray Tube

If a magnetic field is present, in general the electron beam is deflected. But if the beam is parallel or antiparallel to the field, then $\varphi = 0$ in Equation 6.1 and F = 0; there is no force, and hence no deflection. If we find that the electron beam is not deflected when its direction is parallel to a certain axis as in figure 6.5, the B vector must point either up or down along that axis.

If we then turn the tube 90°, φ will be $\pi/2$ in Equation 6.1, and the magnetic force is a maximum; the beam is deflected in a direction perpendicular to the plane of B and v. The direction and magnitude of the deflection determine the direction and magnitude of B. We can perform additional experiments in which the angle between B and v is between 0° and 90° to confirm Equation 6.1 or Equation 6.3 and the accompanying discussion. We note that the electron has a negative charge; the force is opposite in direction to the force on a positive charge.

When a charged particle moves through a region of space where both electric and magnetic fields are present, both fields exert forces on the particle. The total force **F** is the vector sum of the electric and magnetic forces:

$$\mathbf{F} = q(\mathbf{E} + v \times \mathbf{B}) \qquad\qquad \textbf{Equation 6.4}$$

6.5 HOW DO TELEVISIONS WORK?

A few TVs in use today rely on a device known as the **cathode ray tube**, or **CRT**, to display their images. LCDs and plasma displays are more common technologies today. It is even possible to make a television screen out of thousands of ordinary 60-watt light bulbs! You may have seen something like this at an outdoor event like a football game. Let's start with the CRT. (See the Schematic below, Figure 6.6).

The terms **anode** and **cathode** are used in electronics as synonyms for positive and negative terminals. For example, you could refer to the positive terminal of a battery as the anode and the negative terminal as the cathode. In a cathode ray tube, the "cathode" is a heated filament (not unlike the filament in a normal light bulb). The heated filament is in a vacuum created inside a glass "tube." The "ray" is a stream of electrons that naturally pour off a heated cathode into the vacuum.

Since electrons are negative and the anode is positive, it attracts the electrons pouring off the cathode. In a TV's cathode ray tube as shown in Figure 6.6, the stream of electrons are concentrated by a focusing anode into a tight beam and then accelerated by an accelerating anode. This tight, high-speed beam of electrons flies through the vacuum in the tube and then hits the flat screen at the other end of the tube. The screen is coated with phosphor, which glows when struck by the beam.

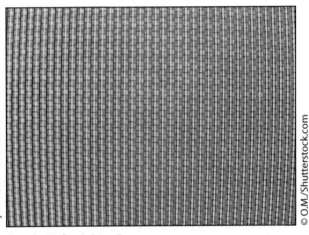

© jultud/Shutterstock.com

© O.M./Shutterstock.com

FIGURE 6.6 Schematic of a Television Set

6.6 MAGNETIC FIELD LINES AND THE FLUX

We can represent any magnetic field-by-magnetic field lines, just as we did for the earth's magnetic field in Figure 6.2. The idea is the same as for the electric field lines. The lines are drawn so that the lines passing through any point is tangent to the magnetic field vector as seen in Figure 6.4 in a lighter shade of grey. In a permanent magnet, the magnetic field lines are drawn so that they pass through the interior of the magnet at each point. The more densely the field lines are packed, the stronger the field is at that point, where they are far apart, the field magnitude is small. (Note: Magnetic field lines are not "lines of force"). Magnetic field lines are sometimes called "magnetic lines of force", but that's not a good name for them. Unlike electric field lines, magnetic field lines do not point in the direction of the force on a charge. Equation 6.2 shows that the force on a moving charged particle is always perpendicular to the magnetic field, and hence to the magnetic field line that passes through the particle's position. The direction of the force depends on the particle's velocity and the sign of its charge, so just looking at magnetic field lines cannot in itself tell you anything about the direction of the force on a moving charged particle. Magnetic field lines do have the direction that a compass needle would point at each location; this may help you to visualize them.

The magnetic flux through a surface can be divided into elements of an area (little pieces). The total magnetic flux through the surface is the sum of the contributions from all of the individual area elements. Magnetic flux is a *scalar* quantity. The SI unit of magnetic flux is equal to the unit of magnetic field (1T) times the unit of area (1 m^2). This unit is called the **weber** (1 Wb), in honor of the German physicist Wilhelm Weber (1804-1891). The total magnetic flux through a closed surface is always zero. Which is symbolically represented as:

$$\iint_s = \mathbf{B} \cdot d\mathbf{A} = 0 \qquad\qquad \textbf{Equation 6.5}$$

This equation is sometimes called *Gauss's law for magnetism*. For Gauss's law, which always deals with *closed* surfaces, the vector area element d**A** (changing area) in the above equation always points out of the surface. However, some applications of magnetic flux involve an open surface with a boundary line; there is then an ambiguity of the sign in Equation 6.5 because of the two possible choices of direction for d**A**. In these cases, we choose one of the two sides of the surface to be the "positive" side and use that choice consistently. If the element of area d**A** is at right angles to the field lines, then the magnitude of the magnetic field is equal to flux per unit area across an area at right angles to the magnetic field. In this case, the magnetic field **B** is sometimes called **magnetic flux density**.

6.7 THE MOTION OF CHARGED PARTICLES IN A MAGNETIC FIELD

When a charged particle moves in a magnetic field, it is acted on by the magnetic force given by Equation 6.2 and Newton's laws determine the motion. Dissecting equation 6.2, a particle with a positive charge q that is moving at right angles to a uniform B field moves in a circle at constant speed due to F and v being perpendicular to each other. In other words, the magnetic force never has a component parallel to the particle's motion, so the magnetic force can never do work on the particle. This is true even if the magnetic field is not uniform. Motion of a charged particle under the action of a magnetic field alone is always motion with constant speed. A real life case of this application would be in a particle accelerator called a cyclotron, where particles moving in nearly circular paths are given a boost twice each revolution, increasing their energy and their orbital radii but not their angular speed or frequency. Similarly, one type of magnetron, a common source of microwave radiation for microwave ovens and radar systems emits radiation with a frequency equal to the frequency of circular motion of electrons in a vacuum chamber between the poles of a magnet.

If the direction of the initial velocity is not perpendicular to the field, the velocity component parallel to the field is constant because there is not a force parallel to the field. Then the particle moves in a helix. Motion of a particle in a non-uniform magnetic field is more complex as in the case of two circular coils separated by some distance. Particles near either coil experience a magnetic force toward the center of the region; particles with appropriate speeds spiral repeatedly from one end of the region to the other and back. Because charged particles can be trapped in such a magnetic field, it is called a *magnetic bottle*. This method is used to confine very hot plasmas with temperatures on the order of 10^6 K. In a similar way the earth's non-uniform magnetic field traps the charged particles, mainly protons and electrons coming from the solar wind in a doughnut-shaped region around earth called the *Van Allen radiation belts* shown in Figure 6.7. They were discovered in 1958 by a group of scientists under the direction of Dr. James Van Allen, thus the name. The group discovered the belts by using data obtained by the instruments on board the Explorer I satellite, which disintegrated in space. The belts are in the upper region of earth's atmosphere called the exosphere. The inner radiation belt extends approximately 600 miles or 1,000 Km above earth's surface and the outer radiation belt is roughly 9,300 miles or 15,000 km to 15,500 miles or 25,000 km above the surface. These charged particles are held by the inner region of earth's magnetic field also known as the magnetosphere (exosphere) which is the light blue region shown below in Figure 6.8 shown as surrounding the Van Allen Belts.

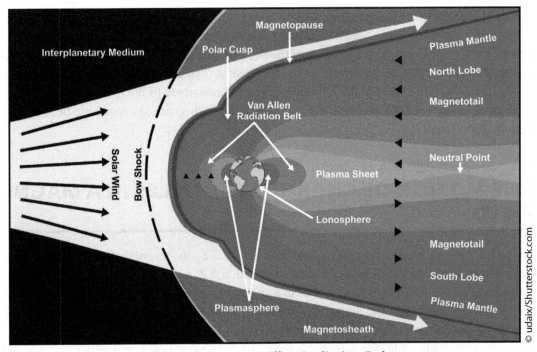

FIGURE 6.7 **Van Allen Radiation Belts**

It is the solar wind that shapes the magnetosphere around earth and other planets. Scientists in NASA's lab have simulated the variable radiation belts on Jupiter as well. What is the relevance of the Magnetosphere? It protects earth against solar flares (violent explosions on the sun releasing large burst of charged particles into the solar system), and the solar wind (a dangerous stream of charged particles coming from the sun). However, there is one benefit, some of these charged particles sneak in the back door of our magnetosphere and interact with the magnetic field lines at the poles creating these beautiful auroras (diffused or discrete). Diffuse auroras

FIGURE 6.8 Earth's Magnetosphere

electrons scatter and the velocities move parallel to the magnetic field lines whereas in discrete auroras the electron's acceleration is aligned with the field. Figure 6.9 is a picture of a discrete aurora taken by a famous photographer, Phil Hoffman in Alaska. The particles collide with the nitrogen and oxygen atoms in the atmosphere and excite those atoms to emit light (photons). In this picture the green seen is the oxygen in the atmosphere. Nitrogen photons would be seen as pink. It takes approximately three days for the charged particles to reach us. These streams of particles amount to as much as ten million megawatts of electrical power or greater. This is enough electrical power to light up a large city. There are also detrimental effects of that much energy being thrown in our direction, the possibility of hitting orbiting satellites. A service such as Spaceweather.com will alert you via phone and email (with a complete copy of the report in your inbox) about the sun's activity. The alerts tell when there is a significant solar flare has been detected and that radiation storms and radio black outs are possible. Since the activity on the sun's surface is variable, this service provides real-time space weather, such as current conditions of solar wind speed, density the class of the solar flare, the number of sun spots per day, a photo gallery and even a program for student participation.

FIGURE 6.9 Aurora Borealis—Bands

6.8 THE APPLICATION OF CHARGED PARTICLES

The motion of charged particles has turned up several applications of the principles introduced in this chapter. One example is the velocity selector; a beam of charged particles produced by a heated cathode or a radioactive material. Particles of a specific speed can be selected from the beam using an arrangement of electric and magnetic fields called a *velocity selector*. Many applications, however, require a beam in which all the particle speeds are the same. Another application was a landmark experiment in physics at the end of the 19th century conducted by an Englishman, Joseph John Thomson (1856–1940) affectionately known as J.J. by his friends and colleagues. He measured the ratio of charge to mass for the electron. This experiment was carried out in 1897 at the Cavendish Laboratory in Cambridge, England. Thomson used an apparatus with a highly evacuated glass container in which electrons are accelerated from the hot cathode and formed into a beam by a potential difference between the two anodes (positive terminals) to measure the ratio of charge to mass. The most significant aspect of his experiment is that he found a single value for this quantity. Thus, Thomson is credited with the discovery of the first subatomic particle, the electron. He also found that the speed of the electrons in the beam was about one-tenth the speed of light, much greater than any previously measured speed of a material particle. Fifteen years after Thomson's experiments, the American physicist Robert Millikan succeeded in measuring the charge of the electron precisely with a value of $m = 9.1093826(16) \times 10^{-31}$ kg. In 1919 a student of Thomson's, Francis Ashton (1877–1945) built the first mass spectrometer.

We have looked at how magnetic fields are created and how they interact with the solar wind, now let's turn our attention to an everyday application: What makes an electric motor work? To answer that question one has to examine the forces that make the motor turn. They are forces that a magnetic field exerts on a conductor carrying a current. The magnetic forces on the moving charges within the conductor are transmitted to the material of the conductor, and the conductor as a whole experiences a force distributed along its length. In a motor, a magnetic torque acts on a current-carrying conductor, and electric energy is converted to mechanical energy. As an example, take a look at a simple type of direct current (DC) motor below. The moving part of the motor

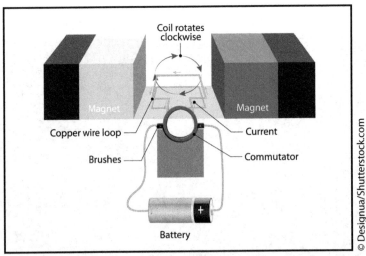

FIGURE 6.10 Diagram of a Simple DC Motor

is the rotor. The rotor is a wire loop that is free to rotate about an axis: the rotor ends are attached to the two curved conductors that form the commutator. The commutator segments are insulated from one another. Each of the two commutator segments makes contact with one of the terminals, or brushes, of an external circuit that includes a source of emf. This causes a current to flow into the rotor on one side, and out of the rotor on the other side. Hence the rotor is a current loop with a magnetic moment. The rotor lies between opposing poles of a permanent magnet, so there is a magnetic field that exerts a torque on the rotor. The torque causes the rotor to turn counterclockwise, in the direction that will align the magnetic moment with the magnetic field.

If the rotor rotates by 90° from its orientation and the current through the rotor is constant, the rotor would now be in its equilibrium state; it would simply oscillate around this orientation. But here's where the commutator comes into play; each brush or terminal is now in contact with both segments of the commutator. There is no potential difference between the commutators, so at this instant no current flows through the rotor, and the magnetic moment is zero. The rotor continues to rotate counterclockwise because of its inertia, and current again flows through the rotor. Thanks to the design of the commutator, the current reverses after every 180° of rotation, so the torque is always in the direction to rotate the rotor counterclockwise. When the motor has come "up to speed", the average magnetic torque is just balanced by an opposing torque due to air resistance, friction in the rotor bearings, and friction between the commutator and brushes. The simple motor has only a single turn of wire in its rotor. In practical motors, the rotor has many turns; this increases the magnetic moment and the torque so that the motor can spin larger loads. Using a stronger magnetic field, which is why many motor designs use electromagnets instead of permanent magnets which can also increase the torque. Electromagnets will be discussed in more detail in chapter seven and while playing with the PhET Sims. This will allow one to become more familiar with how they work and their application. Another drawback of the simple design of the motor is that the magnitude of the torque rises and falls as the rotor spins. This can be remedied by having the rotor include several independent coils of wire oriented at different angles.

Because the motor converts electric energy into mechanical energy or work, it requires electric energy input. If the potential difference between its terminals V_{ab} and the current is I then the power input is $P = V_{ab}I$. Even if the motor coils have negligible resistance, there must be a potential difference between the terminals if P is to be different from zero. This potential difference results principally from the magnetic force exerted on the currents in the conductors of the rotor as they rotate through the magnetic field. This associated electromotive force is called an induced emf; it is also called a back emf because its sense is opposite to that of the current. In chapter 9 we will study induced emf resulting from motion of conductors in magnetic fields.

Loudspeakers are another common application of the magnetic forces on a current-carrying wire as shown in Figure 6.11. We learned in the "Sound Lab" that the volume of a loud speaker can be increased by increasing the amplitude. What happens in the case of the loudspeaker, the radial magnetic field created by the permanent magnet exerts a force on the voice coil that is proportional to the current in the coil; the direction of the force is either to the left or to the right, depending on the direction of the current. The signal from the amplifier causes the current to oscillate in direction and magnitude. The coil and the speaker cone to which it is attached responds by oscillating with the amplitude. Turning up the volume knob on the amplifier increases the current amplitude and hence the amplitudes of the cone's oscillation and of the sound wave produced by the moving cone.

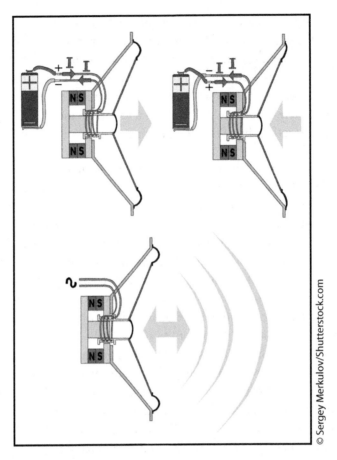

FIGURE 6.11   Loudspeaker components

In this chapter we have studied the forces exerted on moving charges and on current-carrying conductors in a magnetic field. We didn't worry about how the magnetic field got there; we simply took its existence as a given fact. But how are magnetic fields created? We know that both permanent magnets and electric currents in electromagnets create magnetic fields. We've learned that a charge creates an electric field and that an electric field exerts a force on a charge. But a magnetic field exerts a force only on a moving charge. Is it also true that a charge creates a magnetic field only when the charge is moving? In a word, yes.

In the next chapter we will introduce Ampere's law, which plays a role in magnetism analogous to the role of Gauss' law in electrostatics. Ampere's law let us exploit symmetric properties in relating magnetic fields to their sources. Moving charged particles within atoms respond to magnetic fields and can also act as sources of magnetic field. We will use these ideas to understand how certain magnetic materials can be used to intensify magnetic fields as well as why some materials such as iron act as permanent magnets.

GROUP ACTIVITY 5

(This activity assumes that you are already familiar with terms like magnets, magnetic poles, and magnetic fields).

Procedure I

Materials: Magnets and various objects such as, staples, paper clips, nail, aluminum foil, toothpicks, rubber band, key, penny, and dime

Investigate the objects and determine which ones are magnets, interact with the magnets and those that do not interact. Describe the interaction—or lack of interaction—always attracted, always repelled, sometimes attracted, sometimes repelled, or "no interaction."

Are there objects that interact with a magnet but are not magnets?

How would you determine the poles of a magnet that are unlabeled?

Procedure II

Trace a bar magnet. Explore the region around the magnet with a compass. Mark the North and South Pole. Draw arrows indicating the magnetic field at various locations around the magnet. Indicate where the field is strong and weak.

Move the compass away from the objects on your table and shake the compass. What does the needle do? Shake it again, does it point in random positions or in a particular direction?

Explain whether geographic north pole of the Earth is a magnetic north pole or a magnetic south pole.

Look at the PhET simulation "Magnet and Compass" and click on "show field" and "show compass". Describe how these observations compare with Procedure I.

Click on "Show planet Earth". Do your observations agree with what was observed in the previous experiment to the magnetic polarity of the Earth?

1. What is a magnetic field?

2. Explain how magnetic forces and fields interact with a magnet.

3. How would you determine which end of a magnet is the north or south pole?

4. Can a charged particle move through a magnetic field without experiencing any force? If so, how? If not, why not?

5. The magnetic force on a moving charged particle is always perpendicular to the magnetic field **B**, is the trajectory of a moving charged particle always perpendicular to the magnetic field lines? Explain your reasoning.

6. How might a loop of wire carrying a current be used as a compass? Could such a compass distinguish between north and south? Why or why not?

7. How could the direction of a magnetic field be determined by making only qualitative observations of the magnetic force on a straight wire carrying a current?

8. If an emf were produced in a dc motor, would it be possible to use the motor somehow as a generator or source, taking power out of it rather than putting power into it? How might this be done?

9. The magnetic force acting on a charged particle can never do work because at every instant the force is perpendicular to the velocity. The torque exerted by a magnetic field can do work on a current loop when the loop rotates. Explain how these seemingly contradictory statements can be reconciled.

10. Explain how do homing pigeons use Earth's magnetic field as a guidance system.

11. Using the cross product of two vectors (V × B), find the magnitude of the force on a moving charge and its direction if given $V = 5.0 \times 10^7$ m/s, $B = .5T$, $\varphi = 90°$ and $Q = 1.602 \times 10^{-19}$ C.

CHAPTER SUMMARY

The following terminology is used when discussing magnetic forces and fields:

- ▶ **Anode**—positive end of an electrode that attracts electrons pouring off of the cathode.

- ▶ **Cathode**—negative end of an electrode.

- ▶ **Gaussmeter**—instrument used for measuring magnetic fields.

- ▶ **Magnetosphere**—a region in the exosphere surrounding a planet in which charged particles are controlled by the magnetic field.

- ▶ **Magnetic fields**—a moving charge or collection of moving charges (an electric current).

- ▶ **Magnetic forces**—the magnetic force is proportional to the magnitude of the charge, strength of the field, depends on the particle's velocity and is perpendicular to the magnetic field.

- ▶ **Permanent magnets**—do not lose their magnetic field.

© Nicks Stock Store/Shutterstock.com

CHAPTER 7

Electromagnets

The first electrical battery was created by Alessandro Volta in 1799. He named it a voltaic pile. Volta was born into an old Lombard family and began his long illustrious career by making instruments to explore the mysteries of nature. Being somewhat of a self-promoter, he gave public demonstrations of his devices. This discovery was the result of Volta's debate with Luigi Galvani, a professor of anatomy and obstetrics in Bologna, Italy over his famed frog experiment. While performing a lab experiment on a dead frog, Galvani noticed that the frog's leg contracted when touched by a scalpel. Galvani's reasoning was that the jerk of the frog's leg was caused by electricity accumulated in the muscle and traveling through the circuit of the leg. In other words, the frog's leg muscle acted like a charged Leyden jar to activate the nerves of the frog. Volta, a professor of physics at Pavia dismissed Galvani's theory that animal electricity had accumulated in the muscles.

In 1792 Volta was able to successfully duplicate Galvani's experiment and concluded that the twitching of the frog's leg was not because of electrical fluid built up in the muscle but due to the dissimilarity of the two metals used, the scalpel and a metal plate. Volta also concluded that the frog acted in a passive manner but could be thought of as a voltmeter measuring the flow of electricity from one metal to the other. Volta continued perfecting his craft of making instruments and improving upon other scientific devices such as the electroscope and the torpedo fish. After Galvani's death, Volta settled the matter of the frog experiment (animal electricity) versus bimetallic electricity once and for all by introducing the design of a working battery which he never patented, the Voltaic Pile as seen in Figure 7.1. Volta's discovery led to a whole new series of scientific discoveries; electrolysis by William Nicholson and Anthony Carlisle, the electromotive force by Humphry Davy, electricity from voltaic piles by William Wollaston, and the use of voltaic piles in the discovery and research of electric arc effects by Petrov.

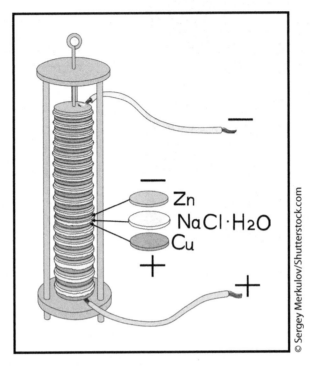

FIGURE 7.1   Volta's Voltaic Pile

It was Michael Faraday who expanded Volta's work. Humphry Davy had done extensive work on electro-magnetism with his mentee Faraday. Faraday capitalized on that work by combining both magnets and the voltaic piles to experiment on electrical devices. He believed that all "electricities" (voltaic, magnetic, thermal, and animal) were all one in the same. Two decades of experimentation had laid the foundation for the electromagnet discovered by William Sturgeon in 1825, Figure 7.2. This picture is the original drawing from Sturgeon's 1824 paper presented to the British Royal Society of Arts, Manufactures, and Commerce. Sturgeon, a British electrical engineer and former soldier was fascinated by science. He taught himself physics and math. In 1824 he became a Lecturer in Science and Philosophy at the East India Company's Military Seminary at Addiscombe, Surrey. The exhibition of his device consisted of a seven-ounce horseshoe shape piece of iron wrapped with several coils of wire and a single battery. It was able to lift nine whole pounds! The power displayed by this device must have generated a lot of excitement in the audience. In 1832 Sturgeon landed another teaching position at the Adelaide Gallery of Practical Science in London and later developed a DC electric motor incorporating a commentator. Before his death in 1850, Sturgeon established with two other associates the London Electrical Society. On his tomb is inscribed, "William Sturgeon—The Electrician". As so often said, "the rest is history"!

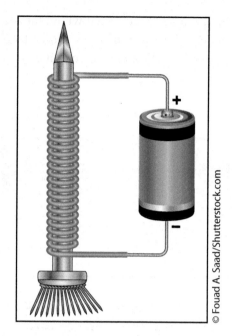

FIGURE 7.2 Sturgeon's Electromagnet Drawing

7.1 WHAT IS AN ELECTROMAGNET?

An electromagnet is a magnet that has electricity running through it. Electrons orbiting the nucleus create magnetic fields in the same way that electrons traveling in the loops of a coil can create a magnetic field. Unlike a permanent magnet, changing the amount of electric current supplied to it, can change the strength of this magnet. When it is turned on, the energy from the circuit is stored in the magnetic field. The magnetic field disappears when the current is turned off and the energy in the field is then returned back to the circuit. Electrons are tiny electromagnets with north and south poles that possess a quality of spin. One can even reverse the poles of an electromagnet by reversing the flow of electricity as seen in Figure 7.3 below. The reason an electromagnet works is because an electric current produces a magnetic field. If a wire carrying an electric current is formed

An electric current flowing towards you produces a magnetic field that circulates in a count-clockwise direction.

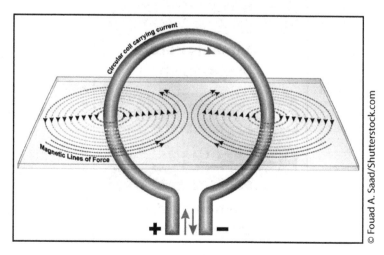

An electric current flowing away from you produces a magnetic field that circulates in a clockwise direction.

FIGURE 7.3 Current flowing through an electromagnet

into a series of loops, the magnetic field can be concentrated within the loops. Wrapping the wire side by side around a core can strengthen the magnetic field even more as shown by Figure 7.4. This forces some of the atoms within the core to point in one direction. The atoms of certain materials, such as iron, nickel, and cobalt, each behave like tiny magnets.

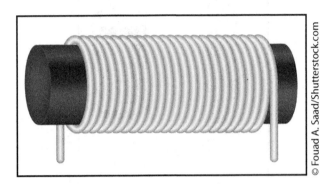

FIGURE 7. 4   **A simple electromagnet consisting of a coil of insulated wire wrapped around an iron core.**

All of their little magnetic fields add together, creating a stronger magnetic field. This results in a magnetic domain, a region of atoms wherein the spins and their resulting magnetic effects are lined up as shown in the bottom of Figure 7.5, domains after magnetization. In some cases pieces of iron in the domains are too small. Therefore, the adjacent domains have fields that point in other directions which implies that the atoms are non-magnetized as illustrated by the top picture of Figure 7.5. This is why non-magnetized iron is preferred over permanent magnets because when the electromagnets are connected to an alternating current source, it has the ability to "flip" the domain. In other words, the direction of the current or magnetic field of the coil can be reversed.

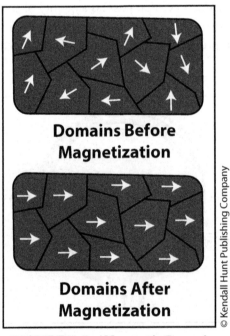

FIGURE 7.5   **Magnetic Domains**

As the current flowing around the core increases, the number of aligned atoms increases, and the stronger the magnetic field becomes as mentioned earlier. Even stronger magnetic fields can be produced if a core is made out of ferromagnetic material. This soft iron is placed inside the coil and the ferromagnetic core increases the magnetic field thousands of times the strength of the field compared to no core as it passes through the center of the coil. As seen in Figure 7.4 above the coil around the core forms the shape of a straight tube or helix known in scientific terms as a solenoid, Figure 7.6. The drawing below shows a cross section through the center of the coil. The circles with a cross in them are wires that show the current is moving into the page while the circles with dots show the wires in which the current is moving up and out of the page. The direction of the magnetic field can be determined by using the right-hand rule. If the fingers of the right hand are curled around the coil in the direction of the current flow (conventional current, flow of positive charge) through the windings, the thumb points in the direction of the field inside the coil. The side of the magnet that the field lines emerge from is defined as the North Pole. Electromagnets have an advantage over permanent magnets in that the magnetic field can be rapidly manipulated over a wide range by controlling the amount of electric current supplied to the system.

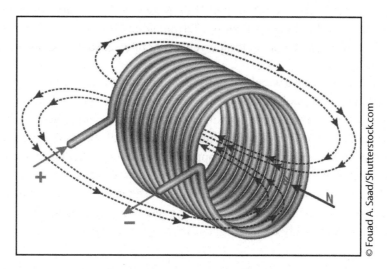

FIGURE 7.6 A solenoid

© Fouad A. Saad/Shutterstock.com

In simulating the PhET, "Magnets and Electromagnets" an electromagnet can be built. To physically build one with an assembled kit, it is fairly easy. All that is needed is to wrap some insulated copper wire around an iron core, attach a battery to the wire, and an electric current will begin to flow and the iron core will become magnetized. When the battery is disconnected, the iron core will lose its magnetism. How can you make your electromagnet stronger? The more turns of insulated wire put on the magnet, the stronger it will be. However, one note of caution, too much current can be dangerous but fortunately, it's safe for the simulation to catch fire! As electricity passes through a wire, some energy is lost as heat. The more current that flows through a wire, the more heat is generated. If you double the current passing through a wire, the heat generated will increase by four times. If you triple it, the heat generated will increase by nine times. As you can see from this example, things can get hot really fast! As you experiment, try different cores and thicker cores just to see what will happen. If a permanent magnet is not attracted to your core, it will not make a very good electromagnet. The factor limiting the strength of electromagnets is the inability to dissipate the enormous waste heat. Using explosives to compress the magnetic field inside an electromagnet as it is pulsed has created the most powerful manmade magnetic fields ever. Although this method sounds destructive, there are definite methods to control the blast so that neither the experiment nor the magnetic structure is harmed. These devices are known as destructive pulsed magnets. They are used in physics and materials science research to study the properties of materials at high magnetic fields.

7.2 HOW ARE ELECTROMAGNETS USED?

Electromagnets are used almost everywhere! Electric motors are a type of electromagnet. Cars have dozens of electromagnets that move things or create electricity. There are other applications for large electromagnets too, such as the dumping of shredded garbage through powerful magnetic fields to sort out and recycle metal bits. At labs around the world scientists shoot particle beams through very powerful magnetic fields and the particles neatly sort themselves out according to their mass, such as the particle accelerator in Switzerland. Electromagnets are widely used in electric and electromechanical devices, such as, motors and generators, transformers, electric bells and buzzers, loudspeakers, earphones, MRI machines, mass spectrometers, particle accelerators and many other modern devices. The most powerful electromagnet in the world, the 45 T hybrid Bitter-superconducting magnet is at the US National High Magnetic Field Laboratory in Tallahassee, Florida. These magnets have a magnetic field higher than the ferromagnetic limit of 1.6 T by using superconducting coils cooled with liquid helium, which conducts a current without electrical resistance.

Have you ever thought about what a wrecking ball, rock concert and your front door have in common? They all use electromagnets. A wrecking yard employs extremely powerful electromagnets to move heavy pieces of scrap metal or even entire cars from one place to another as originally demonstrated by William Sturgeon. Your favorite band uses electromagnets to amplify the sound coming out of its speakers. And when someone rings your doorbell, a tiny electromagnet pulls a metal clapper against a bell. The hardware of most doorbells consists of a metal bell and metal clapper that, when the magnetic charges cause them to clang together, you hear the chime inside and you can answer the door. The bell rings, the guest releases the button, the circuit opens and the doorbell stops its internal ringing. This on-demand magnetism is what makes the electromagnet so useful. With this information we are now ready to investigate the nature of electromagnets!

GROUP ACTIVITY 6

Procedure

A. Simulating Electromagnets: Go to the PhET website and open the simulation "*Magnets and Electromagnets*".

1. Use a DC voltage source, click on "show field," "show compass" and "show electrons". Note the <u>direction</u> and <u>magnitude</u> of the current and the <u>direction</u> and <u>magnitude</u> (strength) of the magnetic field as you vary the voltage and the number of loops.

   The Field Meter indicates the strength of the magnetic field labeling it as "B." List your observations indicating what is varied and how the current and/or field changes and draw diagrams indicating the direction the current is flowing, if possible.

   Use the Left Hand Rule and compare it to the PhET, does it check out?

2. Now switch to AC – Alternating Current

   What are you varying when you move the vertical slider and the horizontal slider?

   Describe what happens when you vary the magnitudes of the current and the frequency?

B. Building a Real Electromagnet

1. To build an electromagnet, a wire, iron nail and batter is needed as illustrated in the figure 7.6 below. Use the Left Hand Rule to determine which end is the North Pole. This can be checked with a compass. Describe what happens once your electromagnet is set up. What happens to the nail?

Courtesy of Gerceida Jones

FIGURE 7.7 **Building an Electromagnet**

1. Do you think increasing the wire coils will (increase, decrease, or not change) the electromagnet's strength?

2. Do you think increasing the voltage will (increase, decrease, or not change) the electromagnet's strength?

3. True/False: Iron is a good metal to use to make an electromagnet.

4. True/False: Magnets and electromagnets are used in many devices.

5. How can you make an electromagnet stronger? Should the core be a permanent magnet or not and why?

6. Where can you find an electromagnet in your home?

7. How can you make a permanent magnet lose its magnetism?

8. The component that produces power in an electromagnetic generator:
   a. armature
   b. field winding
   c. commutator
   d. brush

9. Write about having a "magnetic touch". Think about what you know about magnets. As you are thinking, remember the story of King Midas (a king's wish for everything he touched to turn to gold was granted). Use your imagination and decide which objects you would want to be attracted to you. What would the objects be made of? Would you be able to use the "magnetic touch" to solve problems or would you create more?

10. Look at the following schematic and answer these questions:

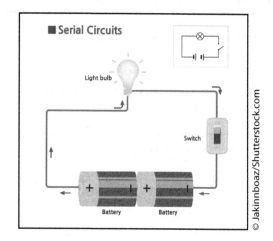

a. When the switch is closed, an electric field will get electrons to flow through the circuit. Look at the wire on the right – rising to the light bulb. State whether the electrons be moving up or down in this wire.

b. Which end of the coil is a North magnetic pole?

c. State two ways in which this electromagnet can be made stronger.

The Face Behind the Hands—Grace Robinson and classmates building an electromagnet.

CHAPTER SUMMARY

The following terminology is used when discussing optics:

► **Circuit**—a closed loop that electrons can travel in. A source of electricity, such as a battery, provides electrical energy in the circuit. Unless the circuit is complete, that is, making a full circle back to the electrical source, no electrons will move.

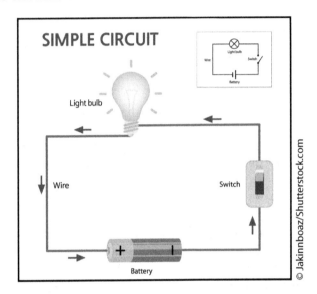

► **Electric current** —a flow of **electric** charge. In electric circuits moving electrons in a wire often carries this charge. It can also be carried by ions in an electrolyte, or by both ions and electrons such as in plasma, i.e. the Sun.

► **Electromagnet**—a soft metal core made into a magnet by the passage of electric current through a coil surrounding it.

► **Ferromagnetic**—of a body or substance) having a high susceptibility to magnetization, the strength of which depends on that of the applied magnetizing field, and that may persist after removal of the applied field. This is the kind of magnetism displayed by iron and is associated with parallel magnetic alignment of neighboring atoms.

► **Right-hand Rule**—is just a convention. The **rule** is also used to determine the direction of the torque vector. If you grip the imaginary axis of rotation of the rotational force so that your fingers point in the direction of the force, then the extended thumb points in the direction of the torque vector.

► **Solenoid**—a cylindrical coil of wire acting as a magnet when carrying electric current.

© Chones/Shutterstock.com

CHAPTER 8

Electricity and Voltage

The history of electricity is a long and convoluted story. It can be traced back long before Thomas Alva Edison started testing his incandescent light bulb. As a matter of fact Edison is not the "Father of Electricity" as so many history books report, he was able to improve on what was already out there by having his lab workers study patents and journals. It was Sir Humphry Davy, an English chemist who was able to make platinum strips flow by passing an electric current through them. The problem with the strips is that they evaporated before they could be made into a useful light source. Two inventors, namely Edison and Swan, would soon solve this impracticality of the filament burning out in a few minutes. In 1848, Joseph Wilson Swan, also an English physicist and chemist began experimenting with incandescent lighting. Edison's model, similar to Swan's design sparked not only a debate but eventually led to a partnership in England called Edison & Swan United Electric Company, Ltd.

Soon inventors from everywhere were trying to create the perfect light bulb; from Warren de la rue (1840), Frederick de Moleyns (1841), Robert Houdin (1851), to Alexander Nikolayevich Lodygin (1872) who received a patent for his incandescent light bulb. However, by 1893, the German inventor Heinrich Gobel claimed to have developed the first carbonized bamboo filament. Across the Atlantic, scientists in Canada and the United States were feverishly working to develop a commercial light bulb. Edison too was experimenting with various filament types. He continued to experiment with various metal filaments. Edison eventually returned to the carbon filaments that led to his first successful test of 13.5 hours. By 1880, Edison improved his design with a carbonized bamboo filament that lasted for over 1,200 hours. One of his team members, William Joseph Hammer, a consulting electrical engineer worked for Edison as a lab assistant on the design of electric lighting.

Another team member, rarely mentioned in the history books is an African American engineer and draftsman, Lewis Latimer.

Lewis Howard Latimer was born in Chelsea, Massachusetts on September 4, 1948. He joined the U.S. Navy at age 15. He later worked as an office boy with a patent law firm. He learned the tools of the trade and was recognized for his talent in sketching patent drawings. He went to work for the U.S. Lighting Company, a rival of Edison. He later received a patent in 1881 for the "Process of Manufacturing Carbons" and also improved on the method of production of carbon filaments used in light bulbs. It was Latimer who created the perfect bulb! He went to work for Edison in 1884 as a draftsman and expert witness in patent litigation on electric lights. Eventually Latimer was inducted into the National Inventors Hall of Fame for his work on electric filament manufacturing techniques. He died December 11, 1928 at the age of 80. His family home is on Latimer Place in Flushing, Queens. He leaves behind a legacy with apartments and schools name in his honor much like Edison who is known as "The Wizard of Menlo Park.

The invention of the electric light bulb did not end just there; soon whole city blocks had lighting from Pearl Street in Manhattan to Chicago's Academy of Music and even some shop windows. However, Edison's goal was to make electric lighting available to the general public. It got to the point where one could rent electric lighting for just about any occasion. Lighting for special events took a bizarre turn in the 19th century, not only were dancers lit with bulbs, electric cocktails were available. All sorts of medical devices were created to cure aliments by passing an electric current through ones body. The market for these fraudulent devices seem to be everywhere with no regulation. This was the twist and turns that led to a revolution in the practical use of electricity. What was the spark of genius that started scientists down this path? Pure and simple, the observation of our natural surroundings! Our study of electricity can begin with static charges.

8.1 ELECTROSTATIC FORCES

We will learn more details about the structure of atoms later, but when atoms arrange themselves in matter sometimes the outermost electrons do not stay attached to any particular atom and are free to wander through the material. Such materials are called *conductors*. There is a large range of conductivity, i.e. electrons move more easily through some materials than through others. Metals such as copper (Cu), silver (Ag), gold (Au), etc. are good conductors. In some other materials, all the electrons are tightly bound to particular atoms and are not free to wander. Some materials can be made to behave sometimes as conductors and sometimes as insulators. These are called *semiconductors* to be discussed later in the book.

Because atoms are overall neutral, material objects are usually overall neutral. However, it is possible to add or subtract electrons and thus make an object negatively charged (electrons added) or positively charged (electrons subtracted). Protons are held in the nucleus by the *strong nuclear force* and it is extremely difficult to remove or add them. We will now begin our study of charges and electric forces.

In the beginning of the semester we read "The Battery" and the chapter in particular on early experiments with amber. This is one way of *charging* objects by rubbing two objects together so that electrons are pulled off of one object (leaving it positively charged) and onto the other object (making it negatively charged). This is what happens when you rub a balloon against, say your sweater and then "stick the balloon to the wall". In the PhET simulation "Balloons & Static Electricity" you can make several observations on the distribution of charges.

8.2   ELECTRIC CHARGES AND ELECTRIC FIELDS

Two charged objects do not have to "touch" in order to attract or repel. The electric force (like the magnetic force) "acts at a distance". An electric charge is surrounded by an *electric field* that sends a message to other charged objects—telling like charges to "go away" and unlike charges to "come here". Similar to magnetic fields, electric fields are represented with arrows. By convention the arrow indicates the direction that a *positive charge* would move if placed in the field. The field around the negative charge tells positive charge "come here" and field around a positive charge tells it to "go away".

The strength of the electric field depends in part on the amount of charge on the object producing that field. Recall that objects get charged when they gain or lose electrons. Therefore, we could measure charge by indicating how many electrons the object gained or lost. However, historically, electric forces and fields were studied before anyone knew that electrons existed. The unit of charge was named a *Coulomb* (C), named for Charles Coulomb who in the eighteenth century studied how the strength of the electric force between two objects depended upon the amount of charge on each object and the distance between them. We now know that a positively (negatively) charged object that has one coulomb of charge has lost (gained) 6.25 billion, billion or 6.25×10^{12} electrons!

8.3   ELECTRICAL ENERGY AND VOLTAGE

You probably already know a bit about energy. For example, you know that if you push on an object, say a wagon, it speeds up. And, as you may have observed, if you continue to push on the wagon, particularly if it is a heavy wagon, you will get tired. You might say that you have "run out of energy". You might have to "refuel" (eat some chocolate) before you can continue pushing! What you would be experiencing is an everyday sense of the Law of Conservation of Energy. Your "lost" energy is the wagon's "gain". The wagon has gained kinetic energy—energy of motion and the amount of kinetic energy an object has depends on its mass and speed. The Law of Conservation of Energy says, "That energy cannot be created or destroyed". More specifically, kinetic energy = ½ mass x speed^2. When mass is measured in kilograms (kg) and speed is measured in meters per second (m/s) then the unit of energy is the unit of mass x speed^2 which is kg(m/s)^2 or a Newton-meter (N-m). This unit has been renamed the *Joule* (N-m), named for James Prescott Joule who in the 1840's did experiments that helped scientists understand energy. Energy can only be changed from one form to another or transferred from one object to another. In the above example, the energy was transferred from you to the wagon.

Now consider the following scenario. You lift an object; hold it briefly, and then let it fall. As it falls, it gains kinetic energy as it moves faster and faster. But since energy cannot be created, that energy had to be there while the object was being held. The raised object stored *gravitational potential energy*. It had the *potential* to fall because it was in a *gravitational* field. The amount of gravitational potential energy stored depends on the mass of the object, how high it is lifted, and the strength of the gravitational field. If the mass is larger or it is lifted higher or the gravitational field strength is stronger than the gravitational potential energy is greater.

Similarly, charged objects can store *electrical potential energy*. For example, if you rub a balloon on your shirt, electrons are pulled off the shirt and on to the balloon. It took energy to pull those electrons away from the positive charges on the atoms in the shirt. This energy came from your rubbing. This energy is now stored as electrical potential energy. Electrical potential energy is stored whenever opposite charges are separated or like charges are brought closer together. In general, consider placing a charged object in an electric field. Upon release, the electric field will exert a force on the object and it will speed up—it will gain kinetic energy and lose electrical potential energy. The amount of electrical potential energy depends on the charge of the object and the strength of the electric field. If the charge is larger or the electric field is stronger the electrical potential energy is greater (See Table 8.1 for definitions below).

Gravitational Potential Energy	Electrical Potential Energy
An object gains *gravitational potential energy* when it is lifted in the presence of a *gravitational field*. The amount of potential energy is proportional to the mass of the object, the strength of the gravitational field and the height to which it is lifted.	An object has *electrical* potential energy when it is in the presence of an *electric field*. The amount of potential energy is proportional to the amount of charge on the object and the strength of the field.

TABLE 8.1 Definitions

We will find it convenient to consider a new quantity that is related to electrical potential energy, namely, *potential energy per charge,* that is, the potential energy one coulomb of charge would have if it were placed in the electric field. For example, if an object with 2 coulombs of charge has 6 joules of electrical potential energy, then one coulomb of charge would have 3 joules of energy. The potential energy per charge = 3 joules/coulomb. The unit "joules/coulomb" has been renamed the *volt* in honor of Alessandro Volta. So, in this example, the *potential energy per charge = 3 volts*. Another name for potential energy per charge is *voltage*—a term that may be more familiar to you. So the next time you look at a 3-volt battery you will know that it gives 3 joules of energy to each coulomb of charge.

8.4 BATTERIES

Suppose you have two oppositely charged metal rods. Excess electrons have been placed on rod A so that it is negatively charged while rod B is deficient in electrons. If a metal wire were connected across from A to B, electrons that are free to move in this conducting wire on rod A will start moving towards rod B. This flow of electrons is called an *electrical current*. And as the electrons "fall" to the positive rod, they lose potential energy. But after a short time all of the electrons will have left the negative rod for the positive, the rods would be neutral and the flow would stop. In order to maintain the flow, it is necessary to maintain negatively and positively charged *terminals*. That is what batteries (and other *voltage sources* such as electrical generators that we will encounter later) do. A battery, via chemical reactions, pulls electrons to higher electrical potential energies – analogous to a water pump lifting water to higher gravitational potential energies. The electrons are then poised to fall through a wire or other device.

8.5   SIMPLE CIRCUITS AND OHM'S LAW

In the "Signal Circuit" activity that you will encounter in this chapter--The flow of electrons (the current) in a simple circuit with only one path (like the one in the "Signal Circuit" simulation) depends on two things: (1) the strength of the electric field in the wire—which depends on the battery or other source of *voltage* and (2) how much *resistance* the electrons encounter as they interact with the atoms in the wire. The amount of resistance differs for different types of wire—the type of material they are made of and how thick and long they are. A better conducting material offers less resistance; a longer wire means more possible interactions and thus more resistance; a thicker wire offers more possible paths to travel and hence less resistance. This is summarized in Ohm's law: Current (I) = voltage (V)/resistance (R) where I is measured in amps, V is measured in volts, and R is measured in *ohms* often written as the Greek letter omega, Ω. Equivalently, multiplying both sides by R:

$$V = I R$$

The relationship between voltage, current, and resistance is often called Ohm's law, but it is important to understand that the real content of Ohm's Law is the direct proportionality (for some materials) of V to I or of the current density to the electric field. The above equation defines resistance for any conductor, whether or not it obeys Ohm's Law, but only when R is constant can we correctly call this relationship Ohm's Law.

Household wiring usually uses a 100-m length of 12-gauge copper wire. At room temperature 25 °C, the wire has a resistance of about 0.5 Ω. A 100-W, 120-V light bulb has a resistance of 140 Ω. If the same current I flows in the copper wire and the light bulb, the potential difference V = IR is much greater across the light bub, and much more potential energy is lost per charge in the light bulb. This lost energy is converted by the light bulb filament into light and heat. You don't want your household wiring to glow white-hot, or else your house will burn down! Therefore, its resistance is kept low by using wire of low resistivity and large cross-sectional area. The resistivity of a material varies with temperature and so does a specific conductor.

8.6   ELECTROMOTIVE FORCE AND CIRCUITS

For a conductor to have a steady current, it must be part of a path that forms a closed loop or complete circuit. Otherwise, there can be no steady motion of charge in an incomplete circuit. To maintain a steady current in a complete circuit, recall a basic fact about electric potential energy: If a charge q goes around a complete circuit and returns to its starting point, the potential energy must be the same at the end of the round trip as at the beginning. There is always a decrease in potential energy when charges move through an ordinary conducting material with resistance. So there must be some part of the circuit in which the potential energy increases; a device called a source of emf, *electromotive force*. This source acts as a water pump, raising a charge from a lower potential energy to a higher potential energy (uphill). Just as a water fountain requires a pump, an electric circuit requires a source of electromotive force to sustain a steady current. Batteries, electric generators, solar cells, and fuel cells are all examples of sources of emf. All such devices convert energy of some form whether it's mechanical, chemical, thermal, etc., into electric potential energy and transfer it into the circuit to which the device is connected. An ideal source of emf maintains a constant potential difference between the terminals, independent of the current through it.

Now let's make a complete circuit by connecting a wire with resistance R to the terminals of a source with the "Circuit Construction Kit" provided by your instructor. We will do further investigations of voltage, current, resistance and Ohm's law in a later activity along with an additional quantity, *electrical power*.

8.7 ELECTRICAL POWER

Let's now look at some energy and power relationships in electric circuits. If you look at a household electrical device, the most prominent electrical information will most likely be *wattage*. Wattage is a unit of *power*, namely how much electrical energy it uses each second. 1 watt = 1 joule/second.

A circuit has a source of electrical energy such as a battery or a generator. This energy gets converted into other forms as the electric field travels through various parts of the circuit causing charged particles to move in particular ways. For example, in a light bulb the motion in the thin filament wire results in heat and light and a 60-watt bulb converts 60 joules of electrical energy into light and heat energy each second. In a speaker, the motion of electrons ultimately results in the energy of a sound wave.

The law of conservation of energy dictates that the electrical energy put into the circuit equal the energy converted into that form. Using the definitions of current and voltage and a little algebra, you can show that in an electrical circuit,

$$\text{Power (P)} = \text{voltage (V)} \times \text{current (I)}$$

$$P = VI.$$

You can check this by checking the units:

1 volt = 1 joule/coulomb.

1 amp = 1 coulomb/second.

Thus, the unit of VI = (1 joule/coulomb) × (1 coulomb/second)

=1 joule/second

=1 watt.

In the case where there is power input to a source; such as the upper circuit element, a car battery, being charged by the car's alternator (the lower element). In other words, work is being done on (the battery), rather than by it. This is what happens when a rechargeable battery (a storage battery) is connected to a charger. The charger supplies electrical energy to the battery; part of it is converted to chemical energy, to be reconverted later, and the remainder is dissipated (wasted) in the battery's internal resistance, warming the battery (as you experienced in building an electromagnet in lab 6) and causing a heat flow out of it. If you own a power tool, laptop computer or cell phone with a rechargeable battery, you may have noticed that it gets warm while charging too. We will start our investigation in this chapter with electrostatic forces and then move into sustained current flow.

GROUP ACTIVITY 7A

Electrostatic Forces and Fields Activity

PROBLEM: TO STUDY CHARGES, ELECTRIC FORCES AND FIELDS

Go to http://phet.colorado.edu and locate the "Balloons & Static Electricity" simulation

Show only one balloon and check the boxes "Show all charges" and "Wall". In observing the PhET describe the charge distribution on the sweater and wall.

Now rub the balloon against the sweater and describe what happens to the balloon and the sweater

Describe what happens when you move the balloon: 1) to the center of the room, close to but not touching the wall and 2) up and down along the wall.

From your observations do you think the balloon is a conductor or an insulator? How about the sweater?

Now open the "Charges and Fields" simulator on the PhET website. There is a box of red positively charged small balls and a box with blue negatively charged balls. Each ball has a charge of $+1nC$, one nanoCoulomb, or 1×10^{-9} Coulomb.

Drag one of the balls from the red box into the simulation area and select 'Show E-field'. What do you observe? What does the depth of color changes represent?

Place a second red ball on top of the first and describe any changes you see. By Adding more red balls do you see a difference?

Click on the 'Clear All' button and drag in one blue ball from the box into the simulation area. Again 'Show E-field' and describe the similarities and differences of the arrangement of the arrows for the charge.

Use the 'Clear all' and start again by placing one $+1nC$ charge and one $-1nC$ charge separated by a distance of about 1 cm and show the field. Sketch the resulting field of the two charges.

Electric Field Hockey Simulation Game

Open the 'Electric Field Hockey' simulation. Try practice mode before you actually play. The game allows you to set up various arrangements of positive and negative charges. Start will level one and continue on to more difficult levels. Requirement: Have fun!

GROUP ACTIVITY 7B

When you walk across a carpet you have probably noticed that if you touch a metal doorknob you get a "SHOCK". Moving charges have transferred from the rug to your body and then discharged onto the metal doorknob. In the next simulation "John Travoltage" you will observe just that. Open the PhET and rub John Travoltage's foot on the carpet and describe what happens.

The small blue circles represent charged particles, are they electrons or protons? Explain. The blue circles spread through John's body why do you think this happened?

Touch John's hand to the doorknob and describe your observations.

Now rub his foot again and again on the rug not touching the doorknob. Keep rubbing until the charges can jump from John's hand to the doorknob. Explain why this happened.

In each case, is John, the doorknob, and the air an insulator or a conductor?

Circuits

Open the simulation "Signal Circuit."

Check the box next to "Show Switch Inside", slide the switch indicator all the way to the right to open.

Check the boxes "Show Electrons", "Show Signal Arrow" (indicate direction *of the force* the electric field exerts on the electrons), and "Paint Electron." The arrows indicate *the* "Paint" (marks an electron next to the switch so you can follow it around the circuit). What happens when the switch is open?

Close the switch; observe the electrons and the signal arrow. Describe your observations of the electrons and the signal arrow. Note: pay close attention

Ohm's Law

Open the PhET simulation "Ohm's Law".

Separately vary the values for voltage and resistance. How does the current change?

Make a table and record four different sets of values for V, R, and I. Design your data to illustrate how the current changes if either V or R changes.

Simple Circuits and Ohm's Law Activity

Series Circuits—Circuits with a single path

Go to http://phet.colorado.edu and locate "Circuit Construction Kit – DC only". Click "Lifelike."

Build a simple circuit that consists of 1 light bulb, and 1 battery Note: the red circles at the end of each circuit must overlap and the light bulb also has TWO circles. To disconnect, you must "right click" if you wish to disconnect.

When the light comes on and the blue dots begin moving your circuit is complete. Make a drawing of your circuit.

Use the tools on the side to get a non-contact ammeter (measures current). Place the ammeter crosshairs over the moving blue dots. Check the current in different locations along the circuit. Then compare the values of the reading. What did you learn from constructing this circuit?

Try these two task: Record the current and measure the voltage across both the battery and the bulb. What value did you get?

SAVE THIS CIRCUIT FOR COMPARISON WITH PARALLEL CIRCUITS

Now build another circuit, this time use two batteries in series. Place the batteries along a line with a wire connecting the positive end of one to the negative end of the other. Do you observe any changes

Measure and record the voltage across the battery combination (+ of one to – of the other) and the light bulb and measure the current.

Do your observations and/or measurements confirm Ohm's law? If so, explain.

Now remove one battery. Place two additional bulbs next to your original light bulb. Are there any changes? If so, compare the previous results to your original one bulb and one battery circuit set-up.

How are these observations related to Ohm's law? If so, explain.

Check the current in different locations and compare those values. Connect the voltmeter across the battery and each light bulb. Make a table and record the voltage and current. Calculate the power in each case.

Parallel circuits provide more than one path for electrons to travel. Design a circuit that has one battery and three light bulbs.

How does the brightness of each light bulb compare with the one in your original series circuit?

For each bulb, connect the voltmeter across the two ends of the bulb and place the non-contact ammeter adjacent to the bulb. Create a table with the three bulbs and one battery; record the voltage, current and power across each bulb. Connect the voltmeter across the battery and place the ammeter adjacent to the battery and enter the values.

What does this data indicate about the circuit?

1. If you walk across a nylon rug and then touch a large metal doorknob, you may get a shock. Why does this tend to happen more on dry days than on humid days?

2. Two metal spheres are hanging from nylon threads. When you bring the spheres close to each other, they tend to attract. Based on this information alone, discuss all the possible ways that the spheres could be charged. Is it possible that after the spheres touch, they will cling together? Explain.

3. In a flashlight, is the amount of current that flows out of the bulb less than, greater than, or equal to the amount of current that flows into the bulb?

4. Can the potential difference between the terminals of a battery ever be opposite in direction to the emf? If it can, give an example. If it cannot, explain why not.

5. Batteries are always labeled with their emf; for instance, an AA flashlight battery is labeled "1.5 volts". Would it also be appropriate to put a label on batteries stating how much current they provide? Why or why not?

6. Current causes the temperature of a real resistor to increase. Why? What effect does this heating have on the resistance? Explain.

7. Why does an electric light bulb nearly always burn out just as you turn on the light, almost never while the light is shining?

8. The energy that can be extracted from a storage battery is always less than the energy that goes into it while it is being charged. Why?

9. Ordinary household electric lines in North America usually operate at 120 volts. Why is this desirable voltage, rather than a value considerably larger or smaller?

10. Water makes life possible: The cells of your body could not function without water in which to dissolve essential biological molecules. What electrical properties of water me it such a good solvent?

CHAPTER SUMMARY

The following terminology is used when discussing optics:

▶ **Conductors**—a closed loop that electrons can travel in. A source of electricity, such as a battery, provides electrical energy in the circuit. Unless the circuit is complete, that is, making a full circle back to the electrical source, no electrons will move.

▶ **Coulomb**—the SI unit of electric charge, equal to the quantity of electricity conveyed in one second by a current of one ampere.

▶ **Electric charge**—is the physical property of matter that causes it to experience a force when close to other electrically charged matter. There are two types of **electric** charges – positive and negative.

▶ **Electric current**—is a flow of an **electric** charge. In an **electric** circuit moving electrons in a wire often carries this charge. It can also be carried by ions in an electrolyte, or by both ions and electrons such as in plasma.

▶ **Electric field**—a region around a charged particle or object within which a force would be exerted on other charged particles or objects.

▶ **Electrical Potential Energy**—or electrostatic potential energy, is a potential energy that results from conservative Coulomb forces and is associated with the configuration of a particular set of point charges within a defined system.

▶ **Electrons**—a stable subatomic particle with a charge of negative electricity, found in all atoms and acting as the primary carrier of electricity in solids.

▶ **Gravitational Potential Energy**—is **energy** an object possesses because of its position in a **gravitational** field. The most common use of **gravitational potential energy** is for an object near the surface of the Earth where the gravitational acceleration can be assumed to be constant at about 9.8 m/s^2.

▶ **Joule**—the SI unit of work or energy, equal to the work done by a force of one newton when its point of application moves one meter in the direction of action of the force, equivalent to one 3600th of a watt-hour.

▶ **Neutron**—a subatomic particle of about the same mass as a proton but without an electric charge, present in all atomic nuclei except those of ordinary hydrogen.

▶ **Power**—the quantity work has to do with a force causing a displacement.

▶ **Proton**—a stable subatomic particle occurring in all atomic nuclei, with a positive electric charge equal in magnitude to that of an electron, but of opposite sign.

▶ **Resistance**—an object is defined as the ratio of voltage across it (V) to current through it (I), while the conductance (G) is the inverse.

▶ **Semiconductor**—a solid substance that has a conductivity between that of an insulator and that of most metals, either due to the addition of an impurity or because of temperature effects. Devices made of semiconductors, notably silicon, are essential components of most electronic circuits.

▶ **Strong Nuclear Force**—In particle physics, the **strong** interaction (also called the **strong force, strong nuclear force, nuclear strong force** or color **force**) is one of the four fundamental interactions of nature, the others being electromagnetism, the weak interaction and gravitation.

▶ **Voltage**—an electromotive force or potential difference expressed in volts.

▶ **Watt**—the SI unit of power, equivalent to one joule per second, corresponds to the power in an electric circuit in which the potential difference is one volt and the current one-ampere.

© Rido/Shutterstock.com

CHAPTER 9

Electromagnetic Induction

There is experimental evidence that a changing magnetic field induces an electromotive force (emf). Every time you swipe your credit card through the card reader, the information coded in a magnetic pattern on the back of your card is transmitted to the cardholder's bank. Why is it necessary to swipe the card rather than hold it motionless in the card reader's slot? This is a question that will be answered in this chapter.

Almost every modern device or machine, from computers to washing machines to power drills, has an electric circuit at the heart of the operation. An electromotive force is required for a current to flow in a circuit. The source of the emf is the battery. But for the vast majority of electrical devices, which are used in your home, dorm, or business; the source is not a battery but an electrical generating station. Such a station produces electric energy by converting other forms of energy: gravitational potential energy (gpe) at a hydroelectric plant, chemical energy from coal or oil-fired plant, or nuclear energy from a nuclear power plant. You might be wondering how this energy conversion is done. In other words, what is the physics behind the production of almost all of our electrical energy needs?

The answer to this known phenomenon is *electromagnetic induction*: If the magnetic flux through a circuit changes, an emf and a current are induced in the circuit. In a power-generating station, magnets move relative to coils of wire to produce a changing magnetic flux in the coils and create an emf, as you will see in the upcoming PhET simulation. Transformers also depend on magnetically induced emfs in the components of their electric power system. Electromagnetic induction is one of the foundations of our technological society.

The central principal in understanding electromagnetic induction comes from *Faraday's law*. This law relates induced emf to changing magnetic flux in any loop, including a closed circuit. We will also discuss Lenz's law, which helps us to predict the direction of the induced emf and current. In this chapter we will learn the principles that would help us understand electrical energy conversion devices such as motors, generators, and transformers. Electromagnetic induction tells us that a time-varying magnetic field can act as a source of the electric field. A time-varying electric field can also act as the source of a magnetic field. These results form a neat package of formulas known as Maxwell's equations that describe the behavior of electric and magnetic fields in any situation.

9.1 INDUCTION EXPERIMENTS

During the 1830's, Joseph Henry carried out several pioneering experiments with magnetically induced emf in England and by Michael Faraday in the United States. We will take a look at several of these experiments now. For example, as you know a stationary magnet does not produce a current. But if you move the magnet in the direction show in Figure 9.1 through the coil or loop, the galvanometer shows a current in the circuit but only while the magnet is moving. If we keep the magnet stationary and move the coil, we again can detect a current during the motion. We call this an *induced current*, and the corresponding emf required to cause this current is called an *induced emf*.

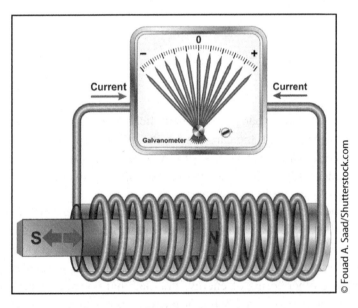

FIGURE 9.1 Demonstrating the Phenomenon of Induced Current

The second example, you can replace the magnet above with another coil but instead connected to a battery. When the second coil is stationary to the first coil there is no current in the first coil. However, when we move the second coil towards or away from the first or move the first towards or away from the second, there is a current in the first coil and the second coil, only if one or the other coil is moving relative to the other.

The final set-up could be to keep both coils stationary and vary the current in the second coil, either by opening and closing a switch or by changing the resistance of the second coil with the switch closed; this will perhaps change the second coil's temperature. We find that as we open or close the switch; there is a momentary current pulse in the first circuit. When we vary the resistance, thus the current, in the second coil, there is an induced current in the first circuit, but only while the current in the second circuit is changing.

9.2 GENERATOR

Let's explore further the common elements in the above examples by looking at a more detailed series experiment as shown in Figure 9.2 below. We connect a coil of wire to a galvanometer, and then place the coil between the poles of an electromagnet whose magnetic field we can vary by a rotating shaft. Here is what we can observe:

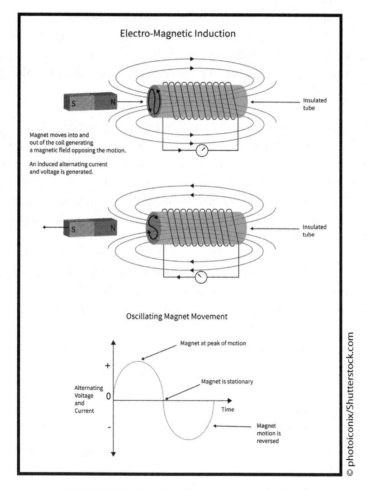

FIGURE 9.2   A coil in a magnetic field

1.   When there is no current in the electromagnet, so that **B = 0**, the galvanometer shows no current.
2.   When the electromagnet is turned on, there is a momentary current through the meter as **B** increases.
3.   When **B** levels off at a steady value, the current drops to zero, no matter how large **B** is.

4. With the coil in a horizontal position, we squeeze it so as to decrease the cross-sectional area of the coil. The meter detects current only during the deformation, not before or after. When we increase the area to return the coil to its original shape, there is current in the opposite direction, but only while the area of the coil is changing.

5. If we rotate the coil a few degrees about a horizontal axis, the meter detects current during the rotation, in the same direction as when we decreased the area. When we rotate the coil back, there is a current in the opposite direction during this rotation.

6. If we jerk the coil out of the magnetic field, there is a current during the motion, in the same direction as when we decreased the area.

7. If we decrease the number of turns in the coil by unwinding one or more turns, there is a current during the unwinding, in the same direction as when we decreased the area. If we wind more turns onto the coil, there is a current in the opposite direction during the winding.

8. When the magnet is turned off, there is a momentary current in the direction opposite to the current when it was turned on.

9. The faster we carry out any of these changes, the greater the current.

10. If all these experiments are repeated with a coil that has the same shape but different material and different resistance, the current in each case is inversely proportional to the total circuit resistance. This shows that the induced emfs that are causing the current do not depend on the material of the coil but only on its shape and the magnet.

The common element in all of these experiments is the changing *magnetic flux* Φ_B through the coil connected to the galvanometer. In each case the flux changes either because the magnetic field is changing with time or because the coil is moving through a non-uniform magnetic field. Check back in the 10 tips above to verify this statement. Faraday's law of induction, the subject of the next section, states that in all of these situations the induced emf is proportional to the rate of change of the magnetic flux Φ_B through the coil. The direction of the induced emf depends on whether the flux is increasing or decreasing. If the flux is constant, there is no induced emf.

Induced emfs are not merely just lab curiosities but have a tremendous number of practical applications. If you are reading these words indoors, you are making use of induced emfs right now! At the power plant that supplied your dorm, an electric generator produces an emf by varying the magnetic flux through coils of wire as shown by Figure 9.2 above. This emf supplies the voltage between the terminals of the wall sockets in your home, office, or dorm room, and this voltage supplies the power to your reading lamp. Indeed, any appliance that you plug into a wall socket makes use of induced emfs.

9.3 FARADAY'S LAW

The common element in all induction effects is changing magnetic flux through a circuit. Faraday's law of induction states: The induced emf in a closed loop equals the negative of the time rate of change of the magnetic flux through the loop. In symbols, it looks like this:

$$\epsilon = \frac{-d\Phi_B}{dt}$$
Equation 9.1

To understand the negative sign, we have to introduce a sign convention for the induced emf ϵ. We can find the direction of the induced emf or current by using equation 9.1 with some simple rules. Here is the procedure:

1. Define a positive direction for the vector area **A** (see Figure 9.3).
2. From the directions of **a** and the magnetic field **B**, determine the sign of the magnetic flux Φ_B and its rate of change $d\Phi_B/dt$ by using the right hand rule.
3. Determine the sign of the induced emf or current. If the flux is increasing $d\Phi_B/dt$, then is positive, then the induced emf or current is negative, if the flux is decreasing, $d\Phi_B/dt$ is negative and the induced emf or current is positive.
4. Finally, determine the direction of the induced emf or current using your right hand. Curl the fingers of your right hand around the A vector, with your right thumb in the direction of A. If the induced emf or current in the circuit is *positive*, it is in the same direction as your curled fingers; if the induced emf or current is *negative*, it is in the opposite direction.

You should check out the signs of the induced emf and currents for the three examples given above. For example, when the loop in Figure 9.2 is in a constant field and we tilt it or squeeze it to *decrease* the flux through it, the induced emf and current are counterclockwise, as seen from above. A word of caution, induced emfs are caused by changes in the flux. Since magnetic flux plays a central role in Faraday's law, it is tempting to think that *flux* is the cause of induced emf and that an induced emf will appear in a circuit whenever there is a magnetic field in the region bordered by the circuit. But Equation 9.1 shows that only changes in flux through a circuit, not the flux itself, can induce an emf in a circuit. If the flux through a circuit has a constant value, whether positive, negative, or zero, there is no induced emf.

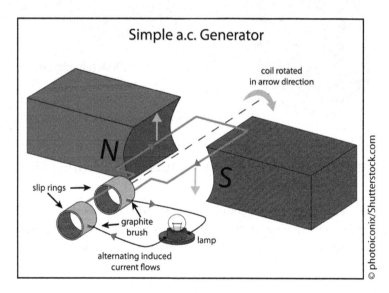

FIGURE 9.3 **Direction of Induced Current or Voltage**

If we have a coil with N identical turns, and if the flux varies at the same rate through each turn, the total rate of change through all the turns is N times as large as for a single turn. If there is a flux through each turn, the total emf in a coil with N turns is:

$$\epsilon = -N\frac{d\Phi_B}{dt}$$
                                                                            Equation 9.2

9.4 LENZ'S LAW

The minus sign in Faraday's law of induction is very important. The minus means that the EMF creates a current I, and a magnetic field B, that oppose the change in flux known as Lenz' law. The direction (given by the minus sign) of the EMF is so important that it is called Lenz' law after the Russian Heinrich Lenz (1804–1865), who, like Faraday and Henry, independently investigated aspects of induction. Faraday was aware of the direction, but Lenz stated it, so he is credited for this discovery (see Figure 9.4). When a bar magnet is thrust into the coil, the strength of the magnetic field increases in the coil. The current induced in the coil creates another field, in the opposite direction of the bar magnet's to oppose the increase. This is one aspect of Lenz's law—induction opposes any change in flux. Verify for yourself that the direction of the induced Bcoil shown indeed opposes the change in flux and that the current direction shown is consistent with the right hand rule in the upcoming PhET simulation in Lab 8.

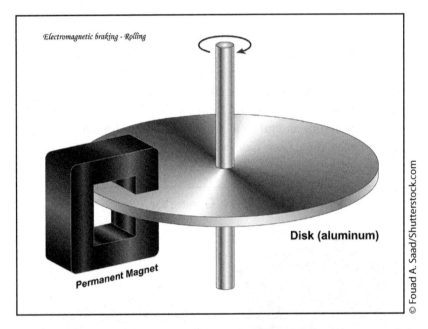

FIGURE 9.4   The Direction of the Induced current using Lenz's Law

As we discussed in this chapter's introduction, induced emfs play an essential role in the generation of electric power for commercial use. Several of the following examples explore different methods of producing emf by the motion of a conductor relative to a magnetic field, giving rise to a changing flux through a circuit, such as a transformer.

9.5 TRANSFORMER

A transformer is a static device, which transforms electrical energy from one circuit to another without any direct electrical connection and with the help of mutual induction between two windings as shown in Figure 9.5 below. This text will be kept short in the interest of time but the general principle behind transformers is simple. The operation of a transformer depends upon Faraday's law of electromagnetic induction as discussed in the previous text. Actually mutual induction between two or more winding is responsible for transformation action in an electrical transformer. If we supply one winding via an alternating electrical source, the alternating current through the winding produces a continually changing flux or alternating flux that surrounds the winding. If any other winding is brought nearer to the previous one, obviously some portion of this flux will link with the second. As this flux is continually changing in its amplitude and direction, there must be a change in flux linkage in the second winding or coil. According to Faraday's Law of Induction there must be an EMF induced in the second. If the circuit of the latter winding is closed, there must be an electric current that flows through it. This is the simplest form of electrical power produced by a transformer.

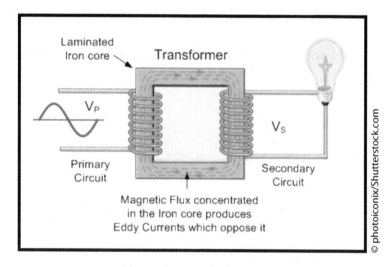

FIGURE 9.5   Electrical Power Transformer

In Lab 8 you will discover how simple it is to induce a current with a permanent bar magnet and a coil of wire shown below from the PhET Sims using *Faraday's Law* that describes how *changing magnetic fields* result in a current. Check it out for yourself!

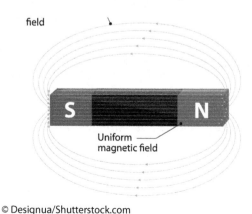

© Designua/Shutterstock.com   © Fouad A. Saad/Shutterstock.com

GROUP ACTIVITY 8

In this group activity students will investigate how *changing magnetic fields* result in a current. The relative strength of the resulting induced current is indicated by the speed of the electron flow, the brightness of the bulb or by the voltmeter as seen in this simulation.

Go to "PhET Simulations" and locate "Faraday's Electromagnetic" lab simulation. Click on the tab called **Pick-up Coil**. Investigate the various ways of using the permanent magnet to induce a current in the coil.

Magnitude

1. Place the magnet to the left of the coil with the North (N) end towards the coil. Both the coil and the magnet are at rest, what do you observe about the current in the coil?
2. Keeping the coil at rest move the N end of the magnet towards the coil. What can you say about the current?
3. Still keeping the coil at rest, vary the speed of the N end of the magnet, what affect does it have on the current?
4. If you bring the N end towards the coil and increase the number of loops in the coil, how does this affect the current?

Direction of the Current

For each of the following, indicate if the current is up or down in the coils.

Action		Current Direction
N →	Coil	
← N	Coil	
S →	Coil	
← S	Coil	
Coil	← N	
Coil	N →	
Coil	← S	
Coil	S →	
N ←	Coil	
N	Coil →	

Note: Think about how you would use a bar magnet and a coil to produce *alternating current (AC)*.

Open the "PhET Simulation" **Transformer** tab. Students will investigate how the two coils can be used to induce a current through the bulb, determining what factors will affect the strength, and direction of the current. With two coils you have to designate which is primary and secondary. The primary one is connected to the battery.

Now choose a setting for the battery. The LHR (left hand rule) is used to determine the polarity of the primary coil. Bring it towards the secondary and note the direction of the induced current. Do your results agree with the similar motion in earlier observations? If so, explain.

Bring the primary close to the secondary. Don't move the primary, but quickly decrease (do not reverse) the voltage and observe the induced current. Again, are there in similarities in the motion of earlier observations?

Generators are devices that convert mechanical energy to electrical energy. Let's investigate this simulation.

Click on the "**Generator**" tab. You will find a water faucet, a compass, a bar magnet on a wheel (turbine), and a coil of wire connected to an incandescent light bulb. Turn on the faucet and describe what you see. Vary the different factors of the simulation and describe what changes you observe and the results of those changes.

1. When a credit card is "swiped" through a card reader, the information coded in a magnetic pattern on the back of the card is transmitted to the cardholder's bank. Why is it necessary to swipe the card rather than holding it motionless in the card reader's slot?

2. If you wiggle a magnet back and forth in your hand, are you generating an electric field? If so, is this electric field? Explain.

3. A sheet of copper is placed between the poles of an electromagnet with the magnetic field perpendicular to the sheet. When the sheet is pulled out a considerable force is required, and the force required increases with speed. Explain.

4. A long, straight conductor passes through the center of a metal ring, perpendicular to its plane. If the current in the conductor increases, is a current induced in the ring? Explain.

5. A circular loop of wire with a radius of 12.0 cm and oriented in the horizontal xy-plane if located in a region of uniform magnetic field. A field of 1.5 T is directed along the positive z-direction, which is upward (draw a picture). (a) If the loop is removed from the field region in a time interval of 2.0 milisecond, find the average emf that will be induced in the wire loop during the extraction process. (b) If the coil is viewed looking down on it from above, is the induced current in the loop clockwise or counterclockwise?

6. Explain the operation of a generator like the one in the simulation. Your explanation should include reference to Lenz's Law and Faraday's Law and include a diagram.

7. Explain the operation of a transformer like the one in the simulation. Your explanation should include reference to Lenz's Law and Faraday's Law and include a diagram.

CHAPTER SUMMARY

The following terminology is used when discussing optics:

- **Faraday's Law of Induction**—the induced emf in a closed loop equals the negative of the time rate of change of the magnetic flux through the loop.

- **Electromagnetic Induction**—is the production of a potential difference (voltage) across a conductor when it is exposed to a varying magnetic field.

- **Galvanometer**—is an instrument used to indicate the presenc'e, direction, or strength of a small electric current.

- **Generator**—is a device that is used to move electrons through a conductor to give electric power. It does this by using a magnet that forces electrons to move along a wire at a steady rate while putting pressure on them.

- **Induced current**—is a current due to variation in the magnetic field surrounding its conductor.

- **Induced emf (electromotive force)**—is the voltage produced by an electric generator in an electrical circuit. The units of measurements are volts.

- **Lenz's Law**—is named after Heinrich Lenz, and it says: An induced electromotive force (emf) always gives rise to a current whose magnetic field opposes the original change in magnetic flux. **Lenz's law** is shown with the negative sign in Faraday's **law** of induction.

- **Magnetic flux**—is the product of the average **magnetic** field times the perpendicular area that it penetrates. It is a quantity of convenience in the statement of Faraday's Law and in the discussion of objects like transformers and solenoids.

- **Transformer**—is an electrical device that transfers energy between two circuits through electromagnetic induction.

© Gordon Ball LRPS/Shutterstock.com

CHAPTER 10

Capacitor and Inductance Coils

The invention of the capacitor depends on whom you talk to. It is thought that the German scientist Ewald Georg Von Kleist first came up with the idea in 1745. However, in 1746 a Dutch scientist named Pieter van Musschenbroek came up with a similar invention, the Leyden jar was discussed in chapter 2 of the "Battery". Although he was the first to build a capacitor, Musschenbroek did not receive credit for it because of a lack of documentation. The Leyden jar is a simple device made of a glass jar, half filled with water and lined inside and out with metal foil and a metal wire or chain driven through a cork on top as seen in Figure 10.1. The chain is hooked to something that would deliver a charge, such as a hand-cranked static generator. The glass jar acts as a dielectric. The jar holds two equal but opposite charges in equilibrium until it is connected with a wire, producing a slight spark or shock.

The American scientist and inventor Benjamin Franklin also worked with the Leyden jar in his experiments with electricity. Franklin discovered that a flat piece of glass worked just as well as the jar model causing him to develop the flat capacitor known as the Franklin Square. But it was Michael Faraday who would pioneer a practical use for the capacitor in storing unused electrons in his experiment. It was Faraday's ingenuity with capacitors that enabled us to deliver electric power over great distances. As a result of his contributions to the field of electricity, the unit of measurement for capacitors (capacitance) became known as the *farad*.

Capacitors have the ability to charge and release their stored charge very quickly allowing them to function in many ways. They hold an important place in everything from voltage stabilizing circuits in sensitive electronics to helping convert AC power to DC. You can even find capacitors in mobility scooters and your laptop computer.

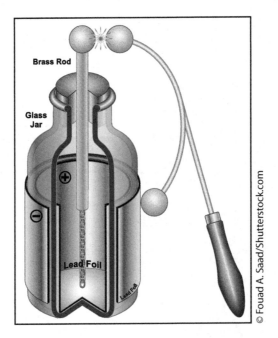

FIGURE 10.1 Leyden Jar

10.1 HOW DOES A CAPACITOR WORK?

When you set an old-fashioned spring mousetrap or pull back the string of an archer's bow, you are storing mechanical energy as elastic potential energy. A capacitor is a device that stores *electric* potential energy and electric charge. In a way, a capacitor is similar to a battery because they both store electrical energy. To make a capacitor, just insulate two conductors from each other. To store energy in this device, transfer the charge from one conductor to the other so that one has a negative charge and the other has an equal amount of positive charge. Work must be done to move the charges through the resulting potential difference between the two conductors, and the work done is stored as electric potential energy. As you know, a battery has two terminals. Inside the battery, chemical reactions produce electrons on one terminal and absorb electrons on the other terminal. However, a capacitor can't produce new electrons, it only stores them. Inside the capacitor the terminals connect two metal plates separated by a non-conducting substance called a dielectric. The plate on the capacitor that is attached to the negative terminal of the battery accepts the electrons (see Figure 10.2) that the battery is producing while the plate on the positive terminal of the battery lose electrons to the battery.

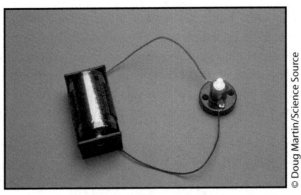

FIGURE 10.2 Capacitor

The dielectric being a non-conductive substance has a lot of practical application. The dielectric dictates the capacitor's use depending on the size and type of dielectric. Some capacitors are better for high frequency uses while other are better for high voltage application, such as glass. Capacitors are manufactured for a variety of daily functions from the smallest plastic capacitor in your calculator, to a Powered-commuter bus. NASA uses glass capacitors to help wake up the space shuttle's circuitry and help to deploy space probes. Specific materials are used to suit the capacitor's function, such as mica, ceramic (high frequency purposes such as antennas, x-ray & MRI machines), porcelain, Mylar (timer circuits for clocks, alarms and counters), Teflon and even air. Believe it or not you can make a simple capacitor from two pieces of aluminum foil and a piece of paper. You can't count on its storage capacity, but it will work!

A capacitor' storage potential (*capacitance*) is measured in units of *farads*. A 1-farad capacitor can store one coulomb of charge at 1 volt. You have already learned that a coulomb is 6.25×10^{18} or 6.25 billion billion electrons. One amp represents a rate of electron flow of one coulomb of electrons per second. Therefore, a 1-farad capacitor can hold 1 amp-second of electrons at 1 volt. A 1-farad capacitor is about the size of a tuna fish can or larger depending on the voltage it can handle. For this reason, capacitors are usually measured in *microfarads*. How big is a farad compared to an AA alkaline battery, which holds about 2.8 amp-hours? If an AA battery can produce 2.8 amps per hour at 1.5 volts, which is about 4.2 watt-hours or it can light a 4-watt bulb for about an hour. Then to store one AA battery's energy in a capacitor, you would need $3,600 \times 2.8 = 10,080$ farads to hold it. That is because an amp-hour is 3,600 amp-seconds. If it takes the size of a tuna fish can to hold a farad, then 10,080 farads is going to take up a lot more space than a single AA battery! The only way to make capacitors useful is to store significant amounts of power at high voltage.

10.2 CAPACITOR APPLICATIONS

Capacitors have a tremendous number of practical applications in devices such as electronic flash units for photography, pulsed lasers, air bag sensors for cars, pacemakers, and radio and television receivers. The difference between a capacitor and a battery is that a capacitor can dump its entire charge in a tiny fraction of a second, where as a battery would take minutes to completely discharge unless it is in severely cold weather. That's why the electronic flash on a camera uses a capacitor instead of a battery. The battery is used to charge up the flash's capacitor over several seconds, and then the capacitor dumps the full charge into the flash tube almost instantly. For this reason it can make a large charged capacitor extremely dangerous. Flash units and TVs have a warning label posted on the device about opening them up. They contain big capacitors that can, potentially, kill you with the charge they contain! A useful definition for this chapter is the word *capacitance*; the ratio of the charge on each conductor to the potential difference between the conductors is a constant as shown by the equation below:

$$C = \frac{Q}{V_{ab}}$$ **Equation 10.1**

The capacitance depends on the sizes and shapes of the conductors and on the insulating material between them, if any. Compared to the case in which there is only vacuum between the conductors, the capacitance increases when an insulating material, such as a *dielectric* is present. This happens because a redistribution of charge, called *polarization*, takes place within the insulating material. In electronic circuits, capacitors are used in several different ways, for instance, to store charge for high-speed use. Capacitors can even-out the voltage of a line carrying a DC voltage that has spikes or ripples in it by absorbing the peaks and filling in the valleys. Finally, a capacitor can block DC voltage too. If you have a situation where you hook a small capacitor to a battery, no current will flow between the poles of a battery once the capacitor charges, shown in Figure 10.3. However, in any alternating current (AC), a signal flows through a capacitor unimpeded. That is because the capacitor will charge and then discharge as the alternating current fluctuates, giving the appearance that the alternating current is flowing. What happens when you connect a capacitor to a battery? Once the battery charges it, the capacitor has the same voltage as the battery (1.5 volts on the battery means 1.5 volts on the capacitor). For a small capacitor, the capacitance is small. But a large capacitor can hold quite a bit of charge as mentioned above. A real life example of a capacitor at work by nature itself is lightning. You can visualize it in this way. One plate is the cloud, the other plate is the ground and the lightning is the charge released between these two "plates." Obviously, in a capacitor that large, you can hold a huge amount of charge!

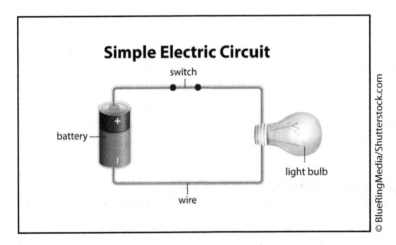

FIGURE 10.3   Configuration of a battery, light bulb and capacitor

The set-up in Figure 10.3 shows a battery, a light bulb, and a capacitor. If the capacitor is pretty big, what you will notice is that, when you connect the battery, the light bulb will light up as the current flows from the battery to the capacitor to charge it up. The bulb will get progressively dimmer and finally go out once the capacitor reaches its capacity. If you then remove the battery and replace it with a wire, a current will flow from one plate of the capacitor to the other. The bulb will light initially and then dim as the capacitor discharges, until it is completely out. One way to visualize the action of a capacitor is to imagine it as a water tower hooked-up to a pipe. A water tower "stores" water pressure – when the water system pump produces more water than a town needs, the excess is stored in the water tower. Then, at times of high demand, the excess water flows out of the tower to keep the pressure up. A capacitor stores electrons in the same way and can then release them for later use.

Capacitors also give us a new way to think about electric potential energy. The energy stored in a charged capacitor is related to the electric field in the space between the conductors. Electric potential energy can be stored in the field itself as shown in Figure 10.4. This figure shows that the electric field lines and equipotential lines for two equal but opposite charges. As you can see the equipotential lines are drawn perpendicular to the electric field lines, if they are known. Note that the potential is greatest (most positive) near the positive charge and least (most negative) near the negative charge. The idea that the electric field is itself a storehouse of energy is at the heart of the theory of electromagnetic waves and our modern understanding of the nature of light. In our opening shot of this chapter, you see a beautiful house with solar panels on its roof. Here is an interesting fact about electromagnetic radiation, if we had the capacity to store just one second of energy from sunlight, we would have enough energy on Earth to last us for the next 500,000 years! Maybe one of you is a visionary. In the next section we will look further into the future uses of capacitors.

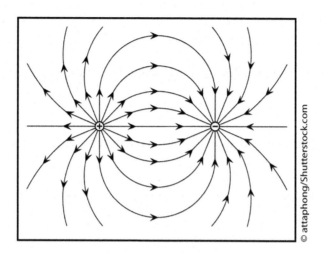

FIGURE 10.4 **The electric field lines and equipotential lines for two equal but opposite charges.**

10.3 FUTURISTIC APPLICATIONS

One recent technology that you are familiar with is touch screens. These are glass screens that have a very thin, transparent metallic coating. A built-in electrode pattern charges the screen so when touched; a current is drawn to the finger and creates a voltage drop. This exact location of the voltage drop is picked up by a controller and transmitted to a computer. These touch screens are commonly found in building directories and more recently in the Apple's iPhone and other similar technology. Another technology you may be familiar with that uses low-inductance high voltage capacitors are particle accelerators, such as the Large Hadron Collider in Geneva, Switzerland. These are groups of large specially constructed capacitor banks (reservoirs). Pulsed lasers are another example of this type of technology. Even nuclear weapons use this type of energy source for the exploding-bridge wire detonators or slapper detonators. Experimental work is under way using banks of capacitors as power sources for an electromagnetic armor, rail guns or coil guns. Reservoir capacitors are used in power supplies where they can smooth the output of a full or half wave rectifier. They can also be used in charge pump circuits as the energy storage element in the generation of higher voltages than the input voltage. One last technology that you may be familiar with is signal processing. The energy stored in this case can be used to represent information, either in binary or analogue form. The problem with high-voltage capacitors is catastrophic failure when subjected to voltages or currents beyond its—ratings or past life expectancy. However, there is a benefit from a pre-charge to high voltage capacitors to limit the rush of currents during the power-up phase of **high voltage direct current (HVDC)** circuits.

Now that you have a little history and the feel for how capacitors work, jump right in and safely experiment with the PhET simulation with no worries.

GROUP ACTIVITY 9

Go to http://phet.colorado.edu and locate the "Capacitor Lab".Click on the first two tabs "Introduction and Dielectric". There is a voltmeter and battery; find variables that are used to design a capacitor.

How would you maximize or minimize to make a capacitor with the greatest capacitance?

Are there any items in the simulation that do not appear to affect the capacitance?

Now explore the relationship between charge, voltage, and stored energy for a capacitor. Summarize your findings.

If you wanted to design a capacitor system to store the greatest amount of energy, what would you use?

Students designed a capacitor below from scratch with the equipment given to them and instructions on how to construct it. It is easy to find a set-up online, just Google "How to build a capacitor".

Courtesy of Gerceida Jones

Photoelectric Experiment Activity

Open the PhET simulation "Photoelectric Effect". When the light source is turned on, the light will hit a target and the current will start to flow.

A resultant "photocurrent" is produced by the ejected electrons as the wavelength and intensity of the light and the type of metal are varied.

Set the battery to zero.

Choose a sodium target and shine red light. Record the wavelength. Slowly increase the *intensity* of the light. What do you observe about the current?

Now very slowly move to shorter and shorter wavelengths until you observe electrons leave the target. Adjust the wavelength slightly to find the maximum wavelength (corresponding to the minimum frequency) of light that results in the ejection of electrons from this target. What is the maximum wavelength?

Keeping the wavelength fixed; increase or decrease the intensity. Now Keep the intensity fixed, slowly move to shorter wavelengths. Observe any changes in the number and speed of the ejected electrons, record your observations.

For a fixed value of wavelength and intensity gradually change the battery voltage so that the right hand plate becomes negative. Record the wavelength and your observations.

At what value of the battery voltage did the current stop (stopping voltage).

Using the same wavelength and a different intensity of light, what can you conclude about the effect of the intensity of light on the stopping voltage?

Now try a different wavelength. What can you conclude about the effect of the wavelength of light on the stopping voltage?

Try different targets and record the maximum wavelength for the photocurrent.

Record your results by answering the following questions: 1) When the wavelength of the light is longer than the maximum wavelength, what is the effect of increasing the intensity? 2) When the wavelength of light is shorter than the maximum, and the intensity is increased, what is the effect on the number and speed of the ejected electrons? 3) As the wavelength of the light is decreased, what is the effect on the number and speed of ejected electrons? and finally 4) What is the main effect of changing the target metal?

CHAPTER SUMMARY

The following terminology is used when discussing capacitors:

- **Capacitive reactance**—The opposition of a capacitor to a sinusoidal current. The unit is the ohm.

- **Capacitor**—An electrical device consisting of two conductive plates separated by an insulating material and possessing the property of capacitance. It also stores electric potential energy and electric charge.

- **Dielectric**—The insulating material between the conductive plates of a capacitor.

- **Electric potential charge**—An *object* may have electric potential energy by virtue of two key elements: its own electric charge and its relative position to other electrically charged *objects*. The term "electric potential energy" is used to describe the potential energy in systems with time-variant electric fields.

- **Farad**—The unit of capacitance.

- **Polarization**—The tendency to be located close to, or attracted towards, one of the two opposite poles of a continuum (Dielectric polarization; charge separation in insulating materials).

1. The capacitance of a capacitor will be larger if
   a. the spacing between the plates is increased
   b. air replaces oil as the dielectric
   c. the area of the plates is increased
   d. all of the above

2. What would happen to the voltage of a capacitor if you increased the charge?

3. The major advantage of a mica capacitor over other types is
   a. they have the largest available capacitances
   b. their voltage rating is very high
   c. they are polarized
   d. all of the above

4. If a 4 farad capacitor stores a charge of 12 coulombs, what is the voltage?

5. Electrolytic capacitors are useful in applications where
   a. a precise value of capacitance is required
   b. low leakage current is required
   c. large capacitance is required
   d. all of the above

6. A switched capacitor emulates a
   a. smaller capacitor
   b. larger capacitor
   c. battery
   d. resistor

7. A pacemaker is surgically implanted in a patient with a malfunctioning sinoatrial node, the part of the heart that generates the electrical signal to trigger heartbeats. To compensate, the pacemaker is located near the collarbone and sends a _____ signal along the lead to the heart to maintain regular beating.
   a. A/C
   b. D/C
   c. Pulsed electrical
   d. Time-varying

8. What happens to the capacitance of a capacitor when the charge on each of the two conductors double?

9. What is the energy stored in a 20 nF parallel plate capacitor when it is charged to 50 volts?

10. What if you had a 9-volt battery charging up a 6 farad capacitor, what is the stored charge?

© Jurik Peter/Shutterstock.com

CHAPTER 11

Quantum Bound States

In quantum physics, the definition of a bound state is a special quantum state for a particle that has potential energy such that the particle has a tendency to stay localized in one or more regions of space. In classical— traditional physics, a bound state is represented by a simple 1-D square well. The potential well is the region surrounding what is known as a local (global) minimum of potential energy. The energy captured in a gravitational potential well does not convert into any other form of energy, such as kinetic energy. In general, the energy spectrum of a set of bound states are discrete and the probability density is finite in the region as time passes. In contrast to the case of free particles in Earth's atmosphere which show a continuous spectrum where the wave packet is spread out as time passes and the probability density at any point tends to zero as time approaches infinity. In elementary chemistry students are taught that electrons when attracted to the nucleus of an atom can inhabit only certain discrete energy states (energy levels) in atoms. In other words, the vast majority of energy levels are in fact forbidden to electrons as in the case of the 1-D model. Therefore, an electron's energy cannot be increased by a small amount; it has to go up in finite—quantum—increments. The study of quantum mechanics is very important in the application of nanotechnology but we will first look at the origin of this science.

11.1 HISTORY OF THE SCHRODINGER EQUATION

In 1926 an Austrian physicist by the name of Erwin Schrodinger, published his time-dependent equation. He described it somewhat like this "The already . . . mentioned psi-function . . . is now the means for predicting probability of measurement results. In it is embodied the momentarily attained sum of theoretically based future expectation, somewhat as laid down in a catalog". Albert Einstein enthusiastically endorsed him who saw, "the matter-waves as an intuitive depiction of nature, as opposed to Heisenberg's matrix mechanics". That same year, the Schrödinger equation detailed the behavior of the wavefunction, ψ but did not mention its nature. Schrödinger tried to interpret it as a charge density in his fourth paper although unsuccessful. It was Max Born who was able to interpret the wavefunction, ψ, as the probability amplitude whose absolute squared is equal to the probability density. In other words, the wave function can be viewed as the displacement of matter. The basis of this equation has become the cornerstone of quantum mechanics. It describes what a system of atoms and subatomic particles will do in the future based on its current state.

The Schrodinger equation itself is very complicated. The solutions are functions describing wave-like motions. In chapter 2 we studied the wave equation for mechanical vibrations on a string. The basis for the Schrödinger's equation is a wavefunction describing the energy of a dynamic system. It is set-up as a linear differential equation based on classical energy conservation, and consistent with the De Broglie relations (wave-particle duality of matter). The solution is the wavefunction Ψ, which contains all the information that can be known about the system itself. The modulus of the wavefunction, Ψ is related to the probability of the particle's spatial configuration at some instant of time. Solving the equation for Ψ can be used to predict how the particles will behave under the influence of the specified potential and with each other.

11.2   ONE-DIMENSIONAL QUANTUM MECHANICS

In the PhET simulation we are observing a finite square also known as a finite potential well. The PhET allows us to imagine a particle confined to a box. The probability vs. position graph shows a peak wavelength of the highest probability of where the electron might be inside of the box. This means that if the total mechanical energy, E^Ψ of the electron is less than the potential energy barrier of the wall $V(x)^\Psi$, the particle will be found inside of the box. The PhET shows a symmetrical 1-dimensional rectangular well of length L on the *x-axis*. We can use the time-independent Schrödinger equation below to describe the energy inside or outside of the box. The simulation describes the wave function behavior in bounded states. The description of basic excitations and their energy spectrums as seen in the 1-D well is of particular interest in modern physics, i.e. nanotechnology research.

The Schrodinger equation looks like this:

$$-\frac{h^2 d^2\psi}{2m\,dx^2} + V(x)\psi = E\psi \qquad (1)$$ **Equation 11.1**

where

$$\hbar = \frac{h}{2\pi},$$

h is Planck's constant

m is the mass of the particle

ψ is the complex value wavefunction that we want to find,

$V(x)$ is a function describing the potential energy at each point x, and

E is the energy of a real number, sometimes called Eigen energy.

Rearranging Equation 11.1:

$$\frac{d^2\psi}{dx^2} = \frac{-2m}{\hbar^2}\left[E - V(x)\right]^\psi (x)$$

Equation 11.2

The conditions put on the wave equation must follow four rules: 1) $\psi(X)$ and $\psi'(X)$ are continuous functions, 2) $\psi(X) = 0$ if X is in a region where it is physically impossible for the particle to be, 3) $\psi(X) \to 0$ as $X \to +\infty$ and $x \to -\infty$, and 4) $\psi(x)$ is a normalized function. Therefore, to solve the Schrodinger Equation, the second order differential equation (11.2), will have two independent solutions) $\psi1(X)$ and) $\psi2(X)$. The general solution will take the form:

$$\psi(X) = A\,\psi1(X) + B\psi2(X)$$

Equation 11.3

Where A and B are coefficients whose values are determined by the boundary conditions. We will not attempt to solve the Schrodinger equation at this time due to the fact that it involves using partial differential equations which is beyond the scope of this course. However, it can be put into a useful form to solve simple problems in this chapter. Now we will return to our discussion of the particle in a 1-D box.

For the case of a particle in a 1-D box of length L, the potential is zero inside of the box at the bottom. However, potential energy rises rapidly as you ascend in the box. The PhET hand shows the amount of energy on each level by simply highlighting the (yellow) horizontal line representing the *x-axis* in equation 11.1. Now the wavefunction shown above is made up of several wavefunctions at different ranges of x depending on whether x is inside or outside of the box. The boundaries of the box are at $x = 0$ and $x = L$. The particles can move freely between those limits at a constant speed and kinetic energy. The regions outside of the box are forbidden. That is the regions where $x<0$ and $x>L$. In other words, the particle cannot leave the box. We know the general solution of the equation $\psi(X) = A\,\psi1(X) + B\,\psi2(X)$ and the boundary conditions at $x = 0$ and $x = L$. Through substitution we end up with an equation in this form:

$$E_n = \frac{n^2\pi^2\hbar^2}{2mL^2}, \hbar = \frac{h}{2\pi}$$

Equation 11.4

if

$$k^2 = \frac{\pi}{L},$$

then the equation turns out to be a well-studied differential equation that can be reduced to a useable form with real numbers shown below as Equation 11.5 by applying suitable boundary conditions.

$$E = \frac{k^2\hbar^2}{2m}.$$

Equation 11.5

Equation 11.4 is the wave functions and probability densities for a particle in a rigid box of length L. The shape (amplitude and wavelength) of the function changes according to the energy levels. For instance, E_n, the allowed energies increase with the square of the quantum number, if n = 1 (the ground-state Energy E_1 is greater than 0). For n = 2, $E_2 = 4E_1$ and n = 3, $E_3 = 9E_1$ etc. According to the Correspondence Principle, when the wavelength becomes small compared to the size of the box, i.e. when either L or the energy of the particle becomes large, the particle must behave as it should. The probability of finding the particle in the region is the fraction of time the particle actually spends in the region. We will not discuss the classically forbidden regions of the 1-D box model.

The Schrodinger equation can be used to solve other more complicated problems such as a finite potential well (quantum well lasers) for a semiconductor diode laser with a single quantum well, a harmonic oscillator, or an electron in a capacitor. A fascinating topic in quantum mechanics is quantum tunneling. This is the idea that electrons or other subatomic particles can "magically" go through a barrier. In this case think in terms of the excess energy gained that is used to cross a barrier. Again, thinking about the laws that governs this process, such as the law of energy conservation and the uncertainty principle are necessary to understand how it works. As ridiculous as this may sound, we already use quantum energy in all of our modern computers and other electronics by controlling the flow rate of quantum tunneling and the electric current by adjusting the height of the energy barrier as just discussed in the 1-D box model.

11.3 QUANTUM THEORY OF LIGHT

The electromagnetic theory of light accounts so well for such a variety of phenomena that it must contain some measure of truth to it, yet this well founded theory is completely at odds with the photoelectric effect. In 1905 Albert Einstein found that the paradox presented by the photoelectric effect could be understood only by carrying further a radical notion proposed five years earlier by the German theoretical physicist Max Planck. Planck was seeking to explain the characteristics of the radiation emitted by bodies of condensed matter. We are all familiar with the glow of a hot piece of metal, which gives off visible light, but other wavelengths to which our eyes do not respond to are present as well. He was able to derive a formula for the spectrum of this radiation (the relative brightness of the various wavelengths present) as a function of the temperature of the radiating body provided; Planck assumed that the radiation is emitted discontinuously in little bursts, an assumption completely at odds with electromagnetic theory. These bursts are called *quanta*. Planck found that the quanta associated with a particular frequency v of light must all have the same energy and that this energy E is directly proportional to v. Therefore, quantum energy is represented by the formula:

$$E = hv \hspace{4cm} \textbf{Equation 11.6}$$

where h, today is known as *Planck's constant* and has a value of 6.626×10^{-34} J-s. While he had to assume that the electromagnetic energy radiated by an object emerges intermittently, Planck did not doubt that it propagates through space as continuous electromagnetic waves. Einstein proposed that light not only is emitted a quantum at a time, but it also propagates as individual quanta, a more drastic break with classical physics. In terms of this hypothesis the photoelectric effect can be readily explained. The empirical formula is:

$$hv = K_{max} + hv_o \hspace{3cm} \textbf{Equation 11.7}$$

Einstein's proposal defined the three terms above in his equation as: 1) hv = energy content of each quantum of the incident light, 2) K_{max} = maximum photoelectron energy and 3) hv_o = minimum energy needed to dislodge an electron from the metal surface being illuminated. There must be a minimum energy required by an electron in order to escape from a metal surface, or else electrons would pour out even in the absence of light. The energy hv_o characteristic of a particular surface is called its *work* function. It states that: Quantum energy

= maximum electron energy + work function of the surface. Some examples of the photoelectric work function are provided in the table below. To detach an electron from a metal surface generally takes about half as much energy as that needed to detach an electron from a free atom of that metal. For instance, the ionization energy of cesium is 3.9 eV compared with its work function of 1.9 eV. Since the visible spectrum extends from about 4.2 to about 7.9×10^{14} Hz, which corresponds to quantum energies of 1.7 to 3.3 eV, it is clear from the table 11.1 below, that the photoelectric effect is a phenomenon of the visible and ultraviolet regions. As shown in the equation above, photons of light whose frequency is v, have the energy hv. To express hv in electronvolts (eV), recall that $1 eV = 1.60 \times 10^{-19}$ J.

TABLE 11.1   Photoelectric Work Functions

Metal	Symbol	Work Function, Ev
Cesium	Cs	1.9
Potassium	K	2.2
Sodium	Na	2.3
Lithium	Li	2.5
Calcium	Ca	3.2
Copper	Cu	4.5
Silver	Ag	4.7
Platinum	Pt	5.6

Hence the formula $E = hv$ now becomes for a single photon energy:

$$E = \frac{6.626 \times 10^{-34} \text{J-s X } v}{1.60 \times 10^{-19} \text{J/eV}} = 4.14 \times 10^{-15} v \text{ eV-s}$$

This equation allows us to find immediately the energy in electronvolts of a photon of frequency v. If we are given the wavelength λ of the light instead, then since $v = c/\lambda$ we have

$$E = \frac{4.14 \times 10^{-15} v \text{ eV-s) X } 3.0 \text{ X } 10^{8} \text{m/s)}}{\lambda} = \frac{1.24 \times 10^{-6} \text{eV-m}}{\lambda}$$

when λ is expressed in meters. When λ is expressed in angstrom units (Å), where $1nm = 10^{-10}$m, then

$$E = \frac{1.24 \times 10^{4} \text{eV-Å}}{\lambda}$$

This is the photon energy. This equation can be used to calculate the maximum kinetic energy of the photoelectrons emitted when light of a particular wavelength falls on a metal surface. Again, this application has extremely important use in such devices as television picture tubes. The emitted electrons evidently obtain their energy from the thermal agitation of the particles constituting the metal, and we should expect that the electrons must acquire a certain minimum energy in order to escape. This minimum energy is always close to the photoelectric work function for the surface.

11.4 BOUNDED STATES IN ASTRONOMY

The stellar evolution of the star determines its fate. The more massive stars will end up as neutron stars and black holes while the lower to intermediate stars will become white dwarfs. White dwarfs, neutron stars, and black holes are the remnants of dead stars. They are a special group of stars known as compact objects (COs). Compact objects no longer burn nuclear fuel. Recall that stars on the main sequence are in hydrostatic equilibrium. This means that the star is balancing gravity, which wants to collapse the star, against thermal pressure, which hold the star up. There are several distinguishing features of these COs, such as physicality, size, and density that will be discussed in this section.

The physicality of a white dwarf (WD) is supported by electron degeneracy pressure. That means that the pressure is caused by the crowding of electrons. This has important implications: 1) it can prevent a collapsing cloud of gas from becoming a star, and 2) in sun-like stars, it controls how they burn helium near the end of their life. The origin of a white dwarf is the evolution of a low to intermediate mass star. Therefore, the size of this compact object is a small radius. The core has a remaining mass less than 1.4 solar masses. This tiny remnant star is approximately the size of Earth upon its demise. If you want to compare how dense this object has become, imagine one sugar cube weighing several tons! What is the fate of the white dwarf, does it just sit there? No, it cools to a black dwarf unless it has a companion. If it has a companion as in the case of Sirius A and B (WD), then Sirius B will steal mass from the external layers of its companion due to the strong gravitational field on the surface of the white dwarf. The accretion of mass causes a thermonuclear explosion of hydrogen gas on the star. As a result, it can come alive again like Lazarus in the Bible! This is known as a nova. If the star gets greedy and stills too much mass, it can't support the weight and the star goes Supernova (Type Ia). This process enriches and returns the gas back into the interstellar medium (ISM).

In the case of a neutron star's physical characteristics, it is formed from very massive stars. It is upheld by neutron degeneracy pressure and is just a ball of neutrons at very high densities making their mass greater than that of an electron. A neutron near the speed of light has high precision. As a result, they can occupy a very small volume of space. The remaining core mass is approximately 1.4 to 3 solar masses and is composed of tightly packed neutrons. These tiny stars are about the diameter of a large city, somewhere around 20 km. The volume or one cubic centimeter (sugar cube) of a neutron star would weigh hundreds of billions of pounds. A better visual, imagine a paper clip having the weight of Mt. Everest! Neutron stars were first detected by Jocelyn Bell; a Cambridge University graduate student. She found a radio source with a regular cycle of exactly 1.3373 seconds. At first scientists thought that she had found an alien civilization. The joke was, Jocelyn has detected the signal of Little Green Men but in fact it was a rapidly spinning neutron star know as a pulsar.

The last example of COs are Black holes. They form from the most massive stars ever and are upheld by neutron degeneracy pressure which can't grow forever. Once the neutron's speed comes close to the speed of light, the pressure fails and gravity has won the fight! Gravity causes the star to shrink even more and the star collapses. The remaining core has a mass more than 3 solar masses. It will continue to collapse into an infinitely small location of space called a singularity. We cannot observe what is left behind but we can detect the presence of one if it has a companion star and attracts the material onto its accretion disk. As you know, it is a collapsed stellar core with enormous gravitational attraction. So strong that not even photons of light can escape it pull. In order to detect a black hole, we look for the x-rays given off by the material falling towards it. Scientists can only speculate on what happens to an object that falls into a black hole. Although it is theorized that space-time is distorted, a classic description is that your feet and head are pulled in the opposite direction. This effect is known as Spaghettification.

11.5 THE APPLICATION OF THE SCHRODINGER EQUATION

Chemists use the equation to describe the electronic energy surfaces on which vibrations, and rotations take place during the motions of electrons and nuclei of molecules. In classical mechanics, the equations play a central role in Newton's laws and the Conservation of Energy. Newton's laws are important as you know because they help us to understand how objects behave when they are standing still, moving, or a force acting upon a body. These laws have universal application. The Conservation of Energy law tells us that the total energy in an isolated system cannot change. In other words, what you put in equals what you get out. And finally, as mentioned earlier, the equations can predict the future behavior of dynamic systems. The modern application of the Schrodinger equation is essential to the use of semiconductors, transistors, and modern computer technology.

Now that you have a feel for what quantum bound states are, you will open the PhET, "Quantum Bound States". Make sure that you keep your "good" eye on those electrons! In group activity 10, we are going to put a lot of atoms together in a periodic array (lattice) creating a solid. The question we want to answer is: what are the allowed states of electrons when we have a lot of atoms strung together. Is it still true that the energy of electrons always has to be increased in finite increments? Run the PhET "Quantum Bound States" and select "Many Wells" tab.

GROUP ACTIVITY 10

Go to http://phet.colorado.edu and locate "Quantum Bound States". Click on the "Many Wells" lab. By stringing together, a lot of atoms in a periodic array, the goal is to create a solid.

Review the energy levels of a single atom. A good example is hydrogen since it is the most abundant element in the Universe:

Choose the following options: 1) Potential Well: Square, 2) Number of Wells: 1, 3) Electric Field: 0.0 V/nm, and 4) Display: Probability Density

The attractive force of the nucleus is shown by the color purple (a *potential well*). The well represents the tendency to fall.

If electrons were not quantum particles, the lowest possible energy state would be for all electrons to fall to the bottom of the well (fall on the nucleus). This does not happen in nature and is not allowed in quantum mechanics.

The graph shown in the PhET simulation is the "probability density" and indicates where an electron will most likely be found.

Horizontal lines represent possible discrete energy levels.

Answer the following questions:

How many energy levels do you observe?

Why is one of the energy levels colored in red?

Now vary the width of the well:

How does the number of possible energy levels change as you increase or decrease the width?

What is the minimal number of energy states you observe?

Now vary the height of the well:

How does the number of possible energy levels change as you increase or decrease the depth?

What is the minimal number of energy states you observe?

Add more atoms changes the number of potential wells (atoms) to two and increase the separation between the wells to the maximum (0.2 nm).

How many energy levels can you see?

Select "Show the magnifying Glass". Look at each energy level with the magnifying glass. What do you observe?

Now vary the height, width, and separation between the wells and record the change in energy levels. Indicate whether you are increasing or decreasing a particular factor.

Change the number of potential wells to five and vary the height, width, and separation between the wells and record the changes you observe in the energy levels. Again, indicate whether you are increasing or decreasing a particular factor.

QUESTIONS AND ANSWERS

1. Based on your observations, what is the effect of increasing the: a) width of the well? b) height of the wells? and c) separation between the wells?

2. How does the separation between adjacent energy levels change as one increases the number.

3. Make a table and summarize your results for the number of energy levels versus number of wells (1,2, 5, and 10).

4. Imagine continuing increasing the number of wells (say, to 10^{23}). Based on your observation, make a sketch of how electron energy levels of 10^{23} of atoms would look like?

5. In what way are the locations of energy levels for 10^{23} atoms are similar to that of one atom?

6. In what way are the locations of energy levels for 10^{23} atoms are different to that of one atom?

7. It turns out that people refer to allowed electron states in solids as bands in distinction to levels in atoms. Why do you think that may make sense?

8. Is it true that the energy of electrons in solids always has to be increased in finite increments? Explain your answer.

9. Find the maximum kinetic energy of the photoelectrons emitted when ultraviolet light of wavelength 3,500 Å falls on the eight surfaces using Table 1.1.

10. Which metal surface would make possible (work best) the operation of such devices as television picture tubes, in which metal filaments or specifically coated cathodes at high temperature supply a dense streams of electrons.

CHAPTER SUMMARY

The following terminology is used when discussing quantum bound states:

- **Black holes** are compact objects that have a bottomless pit in space-time. Nothing can escape a black hole, not even light.

- **Bound States** are a special quantum state of a particle that is subject to a potential.

- **Compact Objects** such as black holes, neutron stars and white dwarfs undergo interactions involving all four the known fundamental forces in nature.

- **Conservation of Energy Law** states that energy can never be created nor destroyed, but can only change form.

- **Dynamical systems concept** is a mathematical formulation which describes the time dependence of a point's position in its ambient space.

- **Energy spectrum** is an arrangement of particle energies in a heterogeneous beam that is analogous to the arrangement of frequencies in an optical spectrum.

- **Isolated System** in the physical sciences are systems removed from and do not interact with other systems.

- **Kinetic energy** is energy in motion. The formula is ½mv².

- **Modulus** is a physical quantity that expresses the degree to which a substance possesses a property, such as elasticity.

- **Neutron degeneracy pressure** is the pressure exerted by neutron.

- **Neutron star** is a compact remnant of a high mass star left over from a supernova explosion.

- **Potential Energy** is the energy stored which can be converted into kinetic energy of motion. There are several forms: 1) gravitational, 2) electrical, and 3) chemical.

- **Potential well** is a region surrounding a local minimum of potential energy.

- **Probability** is defined as something that is probable or the likelihood of something happening.

- **Wavefunction** in quantum physics is a mathematical description of the quantum state of a system.

- **White dwarfs** are compact, hot remnants of a low mass star.

LABORATORIES

HISTORY OF THE UNIVERSE

LAB 1—Estimating the Mass of the Earth

LAB 2—Final Essay Guidelines

LAB 3—Estimating the Mass of the Sun

LAB 4—Atomic Spectra

LAB 5—Classification of Stellar Spectra

LAB 6—Hubble's Law

LAB 7—Visit to the American Museum of Natural History

LAB 1

Estimating the Mass of the Earth

LEARNING OBJECTIVES

▶ To estimate the mass of the Earth using measurements of mass and volume of a rock sample and a geometrically determined value for the volume of the Earth.

▶ To review and use the metric system, exponents, significant digits, and unit conversions when collecting data and performing calculations.

INTRODUCTION

A scientific model is a useful representation of a natural phenomenon or principle that allows us to understand and make predictions about the properties and/or behavior of that system. Models can be used to tackle problems that may seem impossible at first. In today's lab, you will design a simple model to determine the mass of Earth based on a very limited set of measurable quantities.

You will determine the mass of the Earth using only:

1. a rock sample from the surface of the Earth

   ▶ For this model, you will have to assume that the earth is made up entirely of rock of the same density (i.e. the same amount of mass per unit of volume).

2. data that will allow you to calculate the volume of the Earth based on the geometrical relationship that the volume of a sphere is equal to $4/3\pi r^3$

   ▶ For this lab, you will have to assume that the Earth is spherical with a uniform radius of r.

As you might suspect, the value for the mass of the Earth obtained during this lab will be far from accurate. **Consider why.** In Lab 3 you will get a much better value using laws of physics. However, by reconstructing some of the limitations that early astronomers faced when trying to estimate the Earth's mass, this lab will help you appreciate how a fairly simple model can give you a good start at such a seemingly difficult task.

MASS VS. WEIGHT

Measuring the mass of a common object such as an apple is a simple task - it can be placed on a scale designed to measure the force of the Earth's gravity. With a simple spring scale, for example, the extent to which a spring is stretched or compressed is proportional to the force (weight) placed on it. The force on a scale is equal to the mass of the object multiplied by its acceleration due to gravity ($f = ma$). The heavier an object is, the more massive it is. To calculate the weight of an object (force on the scale or $W = mg$), you can use this information to determine the mass.

In this lab you will use a triple beam balance to give you the mass of the object directly by comparison with objects of known masses.

DURING THE LAB

1. **Determine the mass, volume and density of the rock sample.**

   ▶ Select a small rock sample, and determine its mass using a triple beam balance.

      • Record the mass of the rock in grams (g).

      • Since the mass of the Earth will be a large value, convert your measurement to kilograms (1 kg = 1,000 g).

   ▶ Place the rock in a beaker containing a known volume of water and measure how much water is displaced to determine the volume of the rock.

      • Record the volume in mL. (1 mL = 1 cm³).

      • Since the volume of the Earth will be a large value, convert your measurement to km³.

      • 1 mL = 1 cm³

      • 1 m = 100 cm

      • 1 km = 1,000 m

   ▶ Conversions for units of volume can be a little tricky, so here are a couple hints:

      • $1 m^3 = 1 m \times 1 m \times 1 m = 100 cm \times 100 cm \times 100cm = 10^6 cm^3$

      • $1 km^3 = 1 km \times 1 km \times 1 km = 1,000 m \times 1,000 m \times 1,000 m = 10^9 m^3$

   ▶ Calculate the density of the rock sample by dividing its mass by its volume.

2. **Find the Earth's Volume.**

Find the Earth's volume assuming the Earth is a sphere. As you know, a sphere's volume depends on its radius. The problem is now reduced to finding the radius of the Earth. To do this, you will use a method devised over 2,000 years ago to find the circumference of the Earth. Since the circumference of a circle is equal to 2πr, you can use the circumference you determine to solve for the radius.

The Greek philosopher, Eratosthenes, is reported to have made the observation that on the day when sunlight reached the bottom of a deep pit in Syene (Egypt), a sundial's gnomon in Alexandria, **785 km** north of Syene, cast a shadow of **7.2°** (represented as α on the diagram).

Notice on the first diagram below that two parallel lines (1) the path of the sun's rays casting the shadow in Alexandria and (2) the line between the well in Syene and the center of Earth) are intersected by another line (the line down the gnomon to the center of Earth). Use the geometrical relationships of equal angles shown on the second diagram to determine the angle β shown above. The arc of this "sector" (Syene—Earth's center—Alexandria) is 785 km of the total circumference of Earth.

summer solstice (June 21)

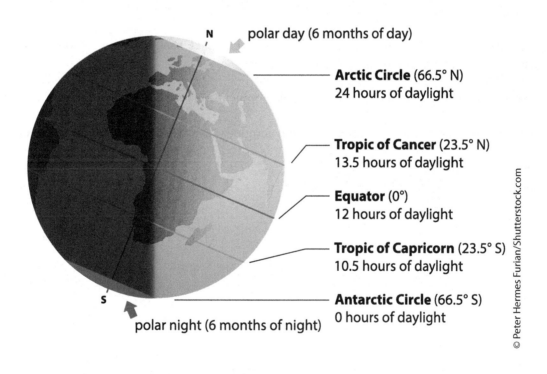

N — polar day (6 months of day)

—— **Arctic Circle** (66.5° N)
24 hours of daylight

—— **Tropic of Cancer** (23.5° N)
13.5 hours of daylight

—— **Equator** (0°)
12 hours of daylight

—— **Tropic of Capricorn** (23.5° S)
10.5 hours of daylight

—— **Antarctic Circle** (66.5° S)
0 hours of daylight

S — polar night (6 months of night)

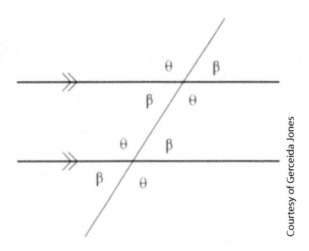

To find the Earth's circumference, Eratosthenes computed how many sectors it would take to fill the Earth's cross-section and multiplied that number by the Syene-Alexandria distance. Estimate the Earth's circumference as Eratosthenes did; remember there are 360° in a circle.

Now, use the circumference to find the radius of the Earth. Then find the Earth's volume using the radius.

3. You have the mass of one rock in kg (m_r) its volume in km³ (V_r) and its density in kg/km³ (ρ_r) You also have the volume of the Earth in km³ (V_E).

4. Devise a method to use the data you have to determine the mass of the Earth.

Describe your method in words and use it to determine the mass of the Earth. Show your calculations.

DATA TABLE AND ANSWER SHEET

Mass of rock sample	g	kg
Volume of rock sample	mL = cm³	km³
Density of rock sample	g/ cm³	kg/ km³
Total # of 7.2° sectors to make 360°	slices	
Circumference of Earth	km	
Radius of Earth	km	
Volume of Earth	km³	**Reminder:** Your estimate for the mass of the Earth should only have as many significant digits as the least significant measurement you used
Estimated Mass of Earth	kg	

Work: Show <u>all</u> your calculations, including conversions, on the next three sheets. You will only receive credit for values in the table above where the work is clearly shown. Your work should be neat and easy to follow.

Use "scientific notation" where appropriate. Indicate the precision of your measurements by the number of significant digits.

Your data/answer sheet and calculations are to be turned in tomorrow.

Additional questions on the Classes website are to be submitted through that website before class tomorrow.

CALCULATIONS: Properties of the rock sample.

Mass of Rock Sample

Grams (Measured) _____

Kilograms: _____

Show calculation:

Volume of Rock Sample

Milliliters = cubic centimeters (Measured) _____

Cubic Kilometers _____

Show calculation

Density of Rock Sample

Grams per cubic centimeter _____

Show calculation

Kilograms per cubic kilometer _____

Show calculation

CALCULATIONS: Volume of the Earth

Number of 7.2 degree slices to make Earth's circumference _____

Show calculation

Circumference of Earth _____

Show calculation

Radius of Earth _____

Show calculation

Volume of Earth _____

Show calculation

CALCULATIONS—Mass of the Earth

Describe your method in words:

Show the calculation you made to determine the mass of the Earth:

Type the answers to the questions assigned by your instructor and show all calculations.

1. Consider the assumptions you made about the composition and shape of the Earth for your model. How do these assumptions affect the accuracy of your answer? Why are you "missing" some of the mass of the Earth?

2. The percentage error is determined by dividing the difference between the actual and calculated masses by the actual mass and multiplying this figure by 100:

    % Error = {(Actual Mass – Calculated Mass) / Actual Mass} × 100

    The actual value for the mass of the Earth is 5.97×10^{24} kg.

    Estimate the % error in your calculated mass of the Earth.

3. Explain the difference between weight and mass.

4. Density equals mass per unit volume ($D = M/V$).

    a) What is the density of the rock sample you used in class in g/cm³?

    b) Calculate the average density of the Earth using the actual values for the Earth's mass (5.9742×10^{24} kg) and volume (1.0832×10^{12} km³). Convert your answer to g/cm³.

    c) Why are your answers for (a) and (b) different?

5. The average radius of Mercury is 2.4×10^3 km. Calculate the volume of Mercury. Express your answer using appropriate significant digits, scientific notation, and adequate units.

LAB 2

Final Essay Guidelines

OVERVIEW

This essay is your opportunity to investigate a current or old question in the History of the Universe and relate it to what we have been learning in our course. You will choose a question from the frequently asked questions in your textbook on pages 167–206. You will write a five to seven (5–7) page paper. This will account for 30% of your final grade.

OBJECTIVES

- ► **We will Meet in Bobst, Room TBA during Lab time (Introduction to Library Sources)**
- ► You have learned about research resources available in NYU's Bobst library, both in the building and via the Internet
- ► Choose the question that you want to investigate
- ► Conduct library and Internet research to gather and analyze information to answer your question
- ► Construct an annotated bibliography of sources organized and formatted using RefWorks
- ► Write a five to seven page paper

OVERALL PROJECT GUIDELINES

This project is composed of four components: annotated bibliography, works cited page, copy of the first page of all references used and the 5–7 page typewritten paper. The graded draft **due date is _____. It will be returned to you approximately three weeks before the term ends and then you must make corrections and bring the original draft and corrected essay with you as part of your final exam.**

You must follow the rubric below in order to receive the maximum number of points. The draft is an opportunity to help you achieve that goal. However, **points will be deducted if the guidelines are not followed (see rubric below)**. Once corrected, you will receive back half of the points lost on the draft. On the day of your final exam, you must return the draft and the final version in order to receive full credit for the essay! No Exceptions!!

The length of the paper is between 5–7 double-spaced pages plus a bibliography (works cited page) included in the paper are parenthetical citations using the MLA style. These will be included as part of your final exam.

The breakdown of points you can earn for the project is in the table below. The information in this table is to help you frame your research.

Section Title	What should be included: Everything in this Rubric including Section Titles (1 point)	Points Available
Introduction	▶ Introduction of topic by listing the main points that you will discuss	1
Background	▶ Basic concepts necessary to understand specifics of topic ▶ Basic vocabulary necessary to understand specifics of topic	2
Recent Work	▶ Recent (< 5 years old) experimental, observational or theoretical work done on your question ▶ What kind of evidence, how was it collected, why/how does it support the answer ▶ Are there any alternate *scientific* explanations or controversies?	6
Importance	▶ Why is this topic important to the general public: economics, legal, religious, cultural? ▶ How will this issue affect people's lives now and in the future? ▶ Are there any new technologies, techniques, treatments that are the result of this work? ▶ How is this related to what we have learned in class?	8
Evaluation	▶ What have you learned that you didn't know before investigating this topic? ▶ What is the "take home message" of your information?	2
Grammar, MLA Format, and Punctuation	▶ Spell checked ▶ Scientific names are in Latin and italicized ▶ Used proper sentence structure, grammar and punctuation ▶ Used proper MLA Format	4
References	▶ At least 3 references: 1 science journal article, 1 magazine/Newspaper, your text book ▶ No commercial web sites, no Wikipedia ▶ All sources are appropriate ▶ Sources cited properly in text of outline ▶ Photocopies of all sources attached	4
Annotated Bibliography	▶ Contains at least 3 references ▶ References are relevant ▶ Each reference has an annotation (https://owl.english.purdue.edu/owl/resource/614/03/), ▶ Used RefWorks	2
	Total Points Available	30

LAB 3

Estimating the Mass of the Sun
(Newton's Version of Kepler's Law)

LEARNING OBJECTIVE

To understand how Newton's version of Kepler's third law is used to determine the mass of the Sun and other central objects in orbiting systems.

INTRODUCTION

A scientific law is a description of a natural phenomenon or principle based on empirical observations. Scientific laws are tested repeatedly and provide generalizations about the behavior of natural systems. Scientific laws accurately predict natural phenomena.

In today's lab, you will use Newton's version Kepler's Third Law of Planetary Motion to calculate the mass of the Sun, Earth and Jupiter.

Kepler's Laws of Planetary Motion

Kepler's excellent mathematical skills and access to Tycho Brahe's accurate planetary position data allowed him to discover three Laws of Planetary Motion.

Kepler's 1st Law of Planetary Motion: The orbits of the planets are ellipses, with the Sun at one focus of the ellipse.

Kepler's 2nd Law of Planetary Motion: The line joining the planet to the Sun sweeps out equal areas in equal times as the planet travels around the ellipse.

Kepler's 3rd Law of Planetary Motion: The ratio of the cube of the distance from the Sun (a^3) to the square of the revolutionary period (P^2) is equal for every planet.

$$\frac{(a_{Venus})^3}{(P_{Venus})^2} = \frac{(a_{Earth})^3}{(P_{Earth})^2} = \frac{(a_{Mars})^3}{(P_{Mars})^2} = \frac{(a_{Jupiter})^3}{(P_{Jupiter})^2} \quad etc.$$

Newton's Law of Universal Gravitation

All matter has an affinity for all other matter. The attraction depends on the amount of matter in each object and on the inverse square of the distance between the objects.

Numerically, we can calculate the gravitational force between two objects using the relationship, $F = -G$ $(m_1 m_2)/d^2$ where m_1 and m_2 are the respective masses of the two objects and d is the distance between them. G is a "units constant" known as the universal gravitation constant. A reasonably precise value, G, was determined experimentally by Henry Cavendish in the century after Newton's death. The value of G was found to be 6.67×10^{-11} m^3/ kg $\times$ s^2.

Using this value of the constant, with units of kilograms for mass, meters for distance and seconds for time, the force will be in (kg $\times$ m/s^2) a standard unit called a Newton [N].

Newton's Cersion of Kepler's Third Law

Kepler's third law can be written as an equation of the form:

$$P^2 = k\, a^3$$

P is the period of a planets orbit and "a" the distance of the planet from the sun (technically, the size of the semi-major axis of the planet's orbital ellipse) and "k" is a "unit constant" to make the numerical answer match the units.

If we use earth years as our unit for period and the average size of the earth's orbit (known as the astronomical unit or AU) for distance then for the case of Earth, by definition, $P = 1$ year, $a = 1$ AU and in these units k, which is 1 year^2/AU^3. Its numerically represented as $P^2 = a^3$. **Note k is equal to 1 only when these units are used.** When we use these units, it is easy to go from a planet's distance from the sun to its orbital period. We will go over this in class.

Newton used his laws of gravity and motion to derive a more accurate version of Kepler's third law. The accurate determination of this law requires calculus. Derivation of a simplified form, using experimental data on the force required to keep an object in circular motion, was shown in class.

In real life, two massive objects will orbit around their common center of mass. Kepler's Third law in this more general case is given to you in the PowerPoint entitled Kepler's Laws.

Newton's more accurate version of Kepler's third law can be written as an equation of the form:

$$(m_1 + m_2)\, P^2 = k\, a^3$$

When the force necessary to keep an object in orbit around a second object is equated to the force of gravity between them, Kepler's Law becomes

$$(m_1 + m_2)\, P^2 = 4\pi^2 a^3 / G;$$

or, rearranging,

$$(m_1 + m_2) = (4\pi^2 / G)\, a^3 / P^2$$

In this equation, m_1 and m_2 represent the masses of the orbiting object. If this mass is in kg, p is in seconds and a in meters, then G, the gravitational constant, 6.67×10^{-11} m^3/kg $\times$ s^2.

If m_1 is much larger than m_2 (as is the case of a planet and the sun and many other cases in the universe) then we can ignore the smaller mass to get the mass of the central object (in this example, the sun). In this case, where the mass of one object is much less than the mass of the other, the mass of the more massive object is given by

$$m = (4\pi^2 / G)\, a^3 / P^2$$

This relationship provides a very useful tool for astronomers. It has been used to determine the masses of the sun, many planets, the galaxy, black holes, clusters of galaxies and many other objects in the universe.

In this lab we will calculate the mass of the Sun and of the Earth using a method similar to that in the PowerPoint 10/12. Please look at that example prior to the lab. As you work the calculations, please don't just crunch the numbers; think about what each number represents.

Once you are familiar with the procedure, we will use a simulated observing program that will allow you to calculate the mass of Jupiter.

PART I   MASS OF THE SUN AND EARTH

Perform all the necessary calculations and unit conversions on a separate sheet of paper. You must show all your work to receive credit for the values in the table below.

Constants Values:

$\pi = 3.14$

$G = 6.67 \times 10^{-11}$ m³/ kg × s²

Solve for the mass of the Sun and Earth using the distances and periods provided. Remember, to make the units work out, you must first get the period of orbit, P, in seconds and the distance between the centers of the objects, a, in meters.

Mass of the Sun

Use the period and distance of Earth

P = 365 days

a = 1.5×10^8 km

Mass of Earth

Use the period and distance of a satellite circling Earth

P = 1.44 hrs

a = 6.5×10^3 km

PART II   MASS OF JUPITER

We will use the CLEA program "The Revolution of the Moons of Jupiter" developed by the Department of Physics at Gettysburg College to simulate observations of the moons of Jupiter and will use that data to determine the mass of Jupiter.

OPEN the program by double clicking on the yellow CLEA icon.

Click FILE and Log in using your initials each lab partners should enter his/her initials—these will show up on the final print out.

Click OK; YES.

Click File; Run; OK.

You will see a screen with Jupiter and its moons.

Point at a moon and left click. The program will identify the moon and give its coordinates with respect to Jupiter in arbitrary units, and, for the X coordinate in Jovian diameters. Find either Europa or Ganymede (they are easier to work with).

We will use the CLEA analysis program so click RECORD, and, in the box that opens, OK.

Click NEXT to get the next day's data. Repeat for about 20 days.

Cloudy night? It happens.

Forget to record your data? In astronomy there are no second chances.

Moon behind Jupiter? That happens too.

That's a real universe out there—be thankful we're not spending 20 cold nights getting this data!

Data Analysis

You will fit a sine curve to your data. This is a good fit for a circular orbit seen edge on.

Click FILE: DATA; REVIEW/EDIT/PRINT

LIST; PRINT (make one copy for each partner) CLOSE

Click FILE; DATA; ANALYSIS

In drop down DATA; SELECT Moon [select your moon]

Click DATA; PLOT; PLOT TYPE; CONNECT POINTS

Click DATA; PLOT; FIT SINE CURVE; SET INITIAL PARAMETERS.

At this point you are giving the computer the basic data from your graph, a starting point, an approximate period and approximate amplitude. The computer will show that approximation on your data plot and you will tweak it to improve the results.

When I tried this with 15 points I got a 0.2% error in the moon's period and a 1.8% error in the size of the orbit.

Using your cursor find a point where your graph crosses the x-axis going from negative to positive. Record this data as T-Zero in the Sine Curve Parameters. You are using a truncated Julian Date. Astronomical dates traditionally were recorded as consecutively numbered days starting in Roman time with the day starting at noon – astronomers worked at night and didn't want the date changing in the middle of their work. The truncation mercifully drops 2,455,000 days from the date.

Using the cursor find the date of the next upward changing zero crossing. Calculate the number of days between the two zeros. This is the tentative period. Record it.

Find the greatest deviation from zero in the plus or minus Y direction. Enter this as amplitude without the plus or minus sign.

Click OK. The computer will fit a sine curve to the data using the parameters you specified.

Use the sliders to improve the fit. You can do it well by eye or you can attempt to minimize the RMS residual.

When you are done, print a copy for each partner.

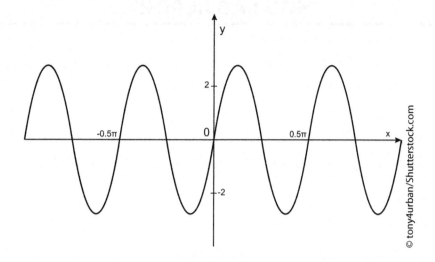

The figure above shows what such a graph will look like although there will be far more data points.

The period fit is the period of the moon in Earth Days.

The amplitude shows the radius of the moon's orbit in Jupiter diameters. Since we know that Jupiter's diameter is 142,984 km we can convert to km (use 143,000 km for the Jovian diameter).

Use the procedure of Part 1 and calculate the mass of Jupiter.

You are the pilot group for this part of the lab. Among other things I want to see if this helps you understand what you are doing and I want to see how the time goes. Can we do two moons in the lab period?

The figures in this lab are from the CLEA lab manual for "The Revolution of the Moons of Jupiter" lab and are used with their permission.

HOU LAB REPORT

Determine the required masses. Show your work, including conversions.

Turn in your data table and graph. Determine the mass of Jupiter. Show your work, including conversions.

ADDITIONAL QUESTIONS

Type the answers to these questions and show all calculations.

Questions directly related to the Lab.

1. Show that Kepler's third law is consistent (approximately) for the orbits of the Mars and Saturn. **Show all your calculations and appropriate units.**

    $P_{Mars} = 690$ days; $a_{Mars} = 2.3 \times 10^8$ km

    $P_{Saturn} = 10,700$ days; $a_{Saturn} = 1.4 \times 10^9$ km

2. The average distance between the Earth and the Sun is called the astronomical unit (AU), which is about 150 million km (1.5×10^8 km). Jupiter is 780 million km from the sun.

    a. How far is Jupiter from the sun in astronomical units (AU)?

    b. Based on your calculation in part a, how many Earth years does it take for Jupiter to orbit the sun in years? **Show all work.**

    (Hint: remember, when using years and AU, $a^3 = P^2$ for all planets)

3. In the late nineteenth century astronomers knew the masses of Mars, Jupiter and Saturn quite accurately. However, the masses of Venus and Mercury were poorly determined. Why was this so?

Questions related to Newton's laws.

4. Explain why it is that a car is experiencing acceleration as it moves around a curve at a constant speed.
5. An insect flying through the air smacks into the windshield of a rapidly-moving train. According to Newton's third law, is the force the windshield exerts on the insect higher than the force the insect exerts on the windshield? Explain.
6. Suppose that you travel to a planet that has 4 times the Earth's mass and 4 times the Earth's radius. Calculate how much more or less you would weigh on this planet compared to your weight on Earth. **Express your answer as a factor or fraction of your weight on Earth. Show all work.**

DATA SHEET 1

Collect 30 lines of data, not counting cloudy nights. Mark cloudy sessions as in the sample below. Remember to enter "E" or "W."

(1) Date	(2) Time	(3) Day	(4) Io	(5) Europa	(6) Ganymede	(7) Calisto
7/24	0.0	1.0	2.95W	2.75W	7.43E	13.15W
7/24	12.0	1.5	CLOUDY–––	–––––––	–––––––	–––––––

(1) Date	(2) Time	(3) Day	(4) Io	(5) Europa	(6) Ganymede	(7) Calisto

DATA SHEET 2

Moon	Period (days)	Period (years)	Semi-major Axis (J.D.)	Semi-major Axis (A.U.)
IV Callisto				
III Ganymede				
II Europa				
I Io				

Recall that 1 year = 365 days, 1 A.U. = 1050 J.D.

Calculating Jupiter's Mass

You now have all the information you need to use Kepler's Third Law to find the mass of Jupiter. But note that values you obtained from the graphs have units of days for p, and J.D. for a. In order to use Kepler's Third Law, you need to convert the period into years by dividing by the number of days in a year (365), and the orbital radius into AU by dividing the number of Jupiter diameters in an AU (1050). Enter your converted value in the spaces provided in **Data Table B** on the previous page. With p and a in correct units, calculate a mass of Jupiter using data from each of your two moons. If one of the values differs significantly from each other, look for a source of error. If no error is found, the data may not be adequate for a better result, in which case you leave the data as you found it.

$$M = \frac{a^3}{p^2}$$

where M_J the mass of Jupiter in units of the solar mass

      a is the radius of the orbit in units of AU

      p is the period of the orbit in Earth years

From Ganymede M_J = _____

From Europa M_J = _____ in units of solar masses

Average M_J = _____ solar masses

NOTE: All values of M_J should be approximately 0.0009 solar masses. This is an approximate value; you should record your own calculated value. If your calculated value seems far from this approximate value, try to give some reason for the discrepancy.

LAB 4

Atomic Spectra

OBJECTIVE

The purpose of this lab is to observe the colors of the visible spectrum, light absorbed and emitted by *excited* atoms and to understand how this light is related to the structure of these atoms.

MATERIALS

- ▶ Spectroscope (A spectroscope is a device that separates light into its component colors.)
- ▶ Various gas discharge tubes and a neodymium light bulb
- ▶ Power supply
- ▶ Spectral chart

WARNING: This lab uses high voltage light sources. DO NOT TOUCH THESE SOURCES AT ANY TIME WHILE THE POWER SOURCE IS ON.

INTRODUCTION
The Colors of the Visible Spectrum

Early investigators of color noticed that white light changes to a <u>spectrum</u> of colored lights after passing through a triangular prism. At first it was thought that the prism "added" something to the white light to make the colors, but in the seventeenth century Sir Isaac Newton demonstrated that this wasn't so. He showed that the white light contained all these colors all along and that all the prism did was separate them. He further showed that the different colors could be recombined to form white light. Newton viewed light as a particle. Later, in the nineteenth century, Young Maxwell, and Einstein proved that light was both a wave and a particle: an electromagnetic wave. It is now known that different wavelengths of light, when they fall on our retinas, cause us to perceive different colors. Although we refer to "the color of light," in reality, "color" applies to our perception of the light and not to the light itself. Light has wavelengths, not colors.

One of the activities you will perform today is to break-up white light into its component colors and learn to identify colors with wavelengths. You can separate white light into its components colors with a device known as a diffraction grating. The colors are spread out different amounts depending on their wavelength and a <u>spectroscope</u> measures the wavelengths. The spectroscope used in this lab will be a self-contained "box" which has wavelengths indicated at the back. These wavelengths are measured in nanometers. One nanometer (nm) equals 0.000000001 meter or 1×10^{-9} m.

PROCEDURE

PART 1

Look through the spectroscope and observe violet, blue, cyan (blue green), green, yellow, orange and red and indicate, approximately, the range of wavelengths for each color you observe. Use the scale below to summarize these observations—or make a list. Note: You and your partners may disagree on some fine points—for example, where orange ends and red begins—so individual records may differ slightly. Record what you see in the box provided in Part 1 on page 189.

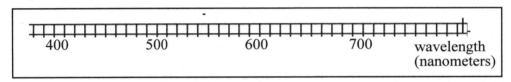

White light

As you have just observed, examples of a **continuous spectra** comes from hot dense bodies, such as, the photosphere of the sun or an incandescent light bulb or even a beautiful rainbow. When scientist first looked closer at the spectrum of sunlight, it appeared to be missing certain colors. These dark or absorption lines in the spectrum remained a mystery until some simple experiments established a connection between the presence of spectral lines and the composition of the source of light. The discovery eventually became known as **absorption spectra**. This breakthrough greatly enhanced our understanding of the natural world.

We are familiar with the structure of atoms – negative electrons orbiting a positively charged nucleus – and we know that different kinds of atoms have different numbers of electrons. However, in order to better understand how individual atoms can emit or absorb light, we must understand the details of how these electrons are arranged.

Electrons can orbit <u>only</u> in certain well-defined paths. The situation is similar to standing on a staircase – you can only stand on the steps, not in between. Similarly electrons can be found only in *allowed orbits*, not in between and each allowed orbit corresponds to a different energy. We know that when a negatively charged electron moves further from the positively charged nucleus it increases its electrical potential energy. That means that electrons in orbits further from the nucleus have higher electrical potential energy than those closer to the nucleus.

Each orbit has a prescribed maximum number of electrons it can hold. When all of the electrons are in the lowest possible orbits, the atom is said to be in the *ground (lowest)* energy state. If an electron is somehow moved to a higher energy orbit, the atom is said to be in an *excited* state. An electron can be raised to a higher energy in three ways:

1. If it is "kicked" there by, say, another electron which knocks into it;
2. If it is "kicked" because the atom collides with another atom;
3. If it absorbs electromagnetic (radiant) energy.

In the gas tubes in this lab, fast-moving electrons move through the tube, "kicking" the gas atom electrons to higher orbits. An excited electron eventually loses its temporarily acquired energy when it returns to a lower energy orbit. It releases this energy as a "packet" of light (or other electromagnetic radiation). This "packet" of light is called a *photon*. Because energy must be conserved, the energy of the photon is equal to the energy lost by the electron as it returns to a lower orbit. Photons of different energies are perceived by us (by our eye/brain system) as different colors. Violet corresponds to the highest energy of visible light and red the lowest. Blue, green, yellow, and orange lie in between, in order of decreasing energy. Ultraviolet light (which we cannot see) and x-rays consist of even higher energy photons than violet light and infra-red, microwaves and electromagnetic radio waves (not to be confused with sound waves) consist of lower energy photons than red light.

When the electron returns to the lowest possible orbit, there may be several ways of making the return. For example, suppose for simplicity, an atom has only three allowed orbits. In the ground state, the electron is in its innermost orbit, labeled 1 in the figure below. If the electron gets "kicked" into orbit 3, it can return to 1 in one large jump or in two small ones as illustrated below. That is, it can emit photons of three possible energies – of three possible colors. If that light were viewed through a standard diffraction grating, like the one used in this lab, you would see three different lines.

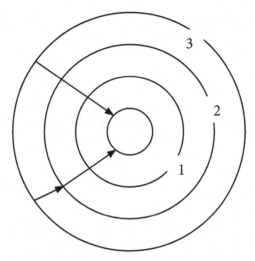

© Kendall Hunt Publishing Company

A. Real atoms have many possible allowed orbits and, therefore, have more than three lines in the spectrum. However, each type of atom has a different arrangement of orbits at different energies so that each type of atom has its own characteristic spectrum. After it was discovered that gases could be made to glow when an electrical current was passed through them, it became relatively easy to identify the spectral lines associated with their presence. The bright lines or **emission** spectra of all known elements have since been cataloged. Emission lines are characteristic of low-density, hot gas emitting light characteristic of the chemical composition of the gas. An emission spectrum is unique to each gas just like fingerprints. Once you have completed the activity, for further investigation and comparison, open the PhET (http://phet.colorado.edu) and the simulation "Neon Lights and Other Discharge Lamps." On the Intro tab click on the "Multiple Atoms" tab. For "Electron Production" choose the "Continuous" button. On the right hand side there is an "Options" tab, check the box for "Spectrometer". Now start the simulation. Compare it to what you observed from your gas tubes.

ABSORPTION SPECTRA

The spectra of the sun and most stars show numerous dark lines against a rainbow background. Why aren't they bright lines as in emission spectra? After all, the Sun is hot – intensely hot! The Sun's visible surface or **photosphere** is in fact too hot for its hydrogen atoms to retain their electrons in any orbit. Atoms such as these are **ionized.** The energy these atoms absorb from the radiation produced deep in the Sun may be released as photons of any wavelength.

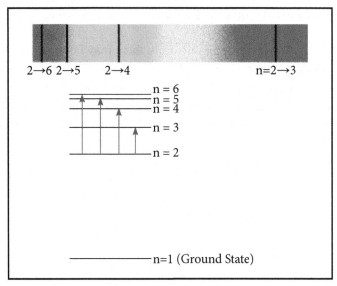

Courtesy of Gerceida Jones

Ordinarily, when we look at the Sun, radiation of all wavelengths passing through its cool **chromosphere** is filtered. Elements in the atmosphere absorb radiation only at wavelengths that allow their electrons to jump to higher orbits according to the Bohr Model. This selective diminution of light at specific wavelengths produces an **absorption spectrum** – dark lines against a continuous background of colors. These lines are characteristic of the composition of the intervening gas – they occur at precisely the same wavelengths as the emission lines produced by the gas at higher temperatures. In this lab you will have the opportunity to observe absorption spectra just as if you were analyzing real starlight. The light source we will use to simulate the sun is a neodymium light bulb and the detector will be your eye.

PROCEDURE

PART II

For each light source assigned by your instructor:

1. Place a gas tube between the contacts of the power supply box.
2. Identify the source and record the colors and corresponding wavelengths. Sketch the wavelengths in the window next to the table as best as you can. If you have problems identifying the wavelengths compare your observations with the spectra on the chart provided.
3. Record your observations in the data table provided on page 189.

PART III

This particular set-up with the neodymium light bulb acts as both the light source and the filter. The light bulb is made with neodymium glass. It filters out any dull, yellow rays that can mask an objects' real color. Once you turn the bulb on, and it begins to heat, neodymium reacts with oxygen to form neodymium oxide (Nd_2O_3), a light blue powder that is made of hygroscopic blue hexagonal crystals. What you are observing is a continuous spectrum emitted by the filament being filtered as light passes through the bulbs own Nd_2O_3 layer (Beehler 2010). You should be able to observe an absorption spectrum. Record your results on the table provided on page 190.

Source: Beehler, Adam J., "Demonstrating spectral band absorption with a neodymium light bulb", The Physics Teacher, Vol. 48, March 2010, page 206.

Answer the following questions on a separate sheet. <u>All answers must be typed</u>.

1. Which has the greatest potential energy, electrons in inner electron orbits or electrons in outer orbits?

2. In your laboratory observations, which of the photons that you observed had the highest energy?

3. How can an electron be made to jump to a higher energy level?

4. What happens when an "excited" electron falls down to a lower energy level?

5. It is often stated that a spectrum is a "chemical fingerprint." What do you think this means?

6. Imagine an emission spectrum produced by a container of hydrogen gas. One of your classmates argues that changing the amount of hydrogen in the container will change the colors of the lines in the spectrum. Is he right? Explain.

7. How do you think scientists might determine what elements are in distant stars?

8. Consider the following diagram of a hypothetical atom. An electron normally orbiting in orbit #2 is "kicked" into orbit #5. Describe the different ways in which it can return to orbit #2. In which way will it emit the higher energy photon?

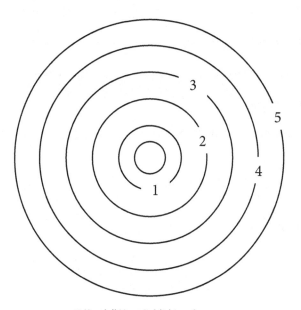

© Kendall Hunt Publishing Company

9. Pick <u>three</u> of the different ways that the electrons can return to orbit #2 in this hypothetical atom. If these resulted in the emission of photons that correspond to red, blue and green, which of the ways would correspond to each of these colors? Indicate the electron transition (e.g. $3 \rightarrow 2$) and the color.

Ophiuchus

ARIES TAURUS GEMINI CANCER

LEO VIRGO LIBRA SCORPIO

SAGITTARIUS CAPRICORN AQUARIUS PISCES

ATOMIC RAP
by Dr. J

Today's lab is the concept of the atomic spectra

Long, long ago the Egyptians adorned the pyramids with electra

Neils Bohr and Rutherford too explained the quirky nature of the electron to you

When these electrons get excited, they jump to another orbit,

A quantum leap it is called

Soon come crashing down to the ground state as they fall,

A photon emitted and absorbed

At just the right frequency of its orb

They disappear, reappear

Not stopping in between

Each element has its own fingerprint as it's seen

Now if for some reason this doesn't make much sense to you
I suggest you study my notes and the New Universe too!

(musical **accompaniment**; 50 Cents—Just a Lil Bit)

LAB 5

Classification of Stellar Spectra

This laboratory exercise is adapted with permission from *The Classification of Stellar Spectra Student Manual:* A Manual to Accompany Software for the Introductory Astronomy Lab Exercise; Document SM 6: Version 1.1.1 lab

OBJECTIVES

► To classify and take spectra of various stars using a simulated telescope and spectrometer.
► To be able to sketch spectral curves of stars including the absorption features and peak intensities drawn approximately at the correct wavelengths.
► Using the Hertzsprung-Russell (H-R) diagram, rank stars according to temperature and luminosity.
► Explore the Chandra X-ray Observatory program, which is available to use in real-time data on your laptops. **Appendix E**

INTRODUCTION

Stars are elegant aggregations of mostly hydrogen and helium atoms held together by a delicate balance between thermonuclear reactions and gravitational forces. They convert a star's mass into energy through the fusion of elements deep in their cores. In 1939, Hans Bethe calculated and explained in a paper how stars shine. "He concluded that stars significantly heavier than the Sun would shine via the CNO cycle and that lighter stars would shine via fusion initiated by the p-p reaction. It laid the conceptual foundation for solving the energy-production problem in main-sequence stars."[1]

Stars have life cycles just as people go through stages in their lives. Eventually, both stars and people will deplete their initial fuel sources. A star's evolution consists of many different stages with fuel consumption as the dominant life cycle of an evolving star. The time scales of stellar evolution depend on the mass of the star. The more mass present, the faster the evolution for the star through the fuel consumption stages. Luminosity (absolute magnitude) one of the star's properties is closely linked with its mass and evolution. Finally, some will explode, enrich the universe with new elements and trigger the formation of new stars, galaxies, and even planetary systems. Astronomers have developed methods to conceptualize visually the mass-luminosity relationship as it pertains to the fuel consumption evolution of stars and plot the different stages in the evolving life cycles of stars. We will look at methods of spectral classification in this lab. **See Appendix A for a historical time line of classifying stars as discussed by Dr. Jones in Lecture.**

PART I

The computer program you will use consists of two parts. **Part I** is a spectrum display and classification tool. This tool enables you to display a spectrum of a star and compare it with the spectra of standard stars in an Atlas of known spectral types. The tool makes it easy to measure the wavelengths and intensities of spectral lines and provides a list of the wavelengths of known spectral lines to help you identify spectral features and to associate them with particular chemical elements. **Please see the Appendix B for further instructions.**

PART II

Part II of the computer program is a realistic simulation of an astronomical spectrometer attached to a research telescope. A TV camera is attached to the telescope so you may see the star's field as shown by **Figure 1** below, and you can view the fields at high and low magnification. You can steer the telescope so that light from a star will pass into the slit of the spectrometer and then turn on the spectrometer and begin to collect photons. The spectrometer display will show the spectrum of the source as it builds up while you collect additional photons. The spectrum is a record of the intensity of the light collected versus the wavelength. When a sufficient number of photons are collected, you should be able to see the distinct spectral lines that will enable you to classify the spectrum. You can use the telescope to obtain spectra for stars. You will then classify your spectra by comparing them with the spectra of standard stars stored in the computer, just as you did in the first part of the exercise. Finally, you will be able to sketch the spectral curves of the stars including the absorption lines and peak intensities at specific wavelengths.

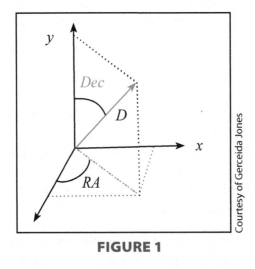

FIGURE 1

The two most important features of a star's blackbody curve are: 1) its maximum height or peak representative of its energy output, and 2) the wavelength at which this peak occurs (peak wavelength). This is an indication of the star's temperature (the longer the peak wavelength, the cooler the star). Take a look at **Figure 3a**; if Star A and B are the same size and temperature, they will have identical blackbody curves. If Star B is the same size as Star A, but cooler, its energy output is less at all wavelengths and the peak occurs at longer wavelengths towards the red end of the spectrum. Viewing again, you should be able to answer these three simple questions: Which star gives off more red light? Blue light? Which star looks redder?

Types of Spectra

We can now distinguish between three types of spectra: 1) Emission Spectrum when light is emitted directly from a low density cloud and passes through a prism, 2) Absorption Spectrum when light is emitted directly from a hot, dense energy source, passes through a low density cloud, and through a prism, and 3) Continuous Spectrum when a hot, dense energy source and passes directly through a prism. Stars like our Sun have low density, gaseous atmospheres surrounding their hot, dense cores. If you were looking at the Sun's or any star's spectrum, which of the three type named above would you observe?

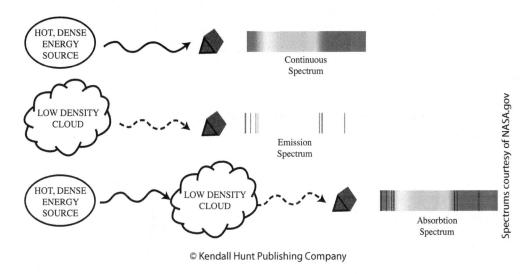

© Kendall Hunt Publishing Company

We are now armed with enough information to begin analyzing a star's spectra. Remember that scientists recognize the chemical differences among stars but the main thing that determined the spectral type of a star was its surface temperature.

PART IIIA

This exercise will require you to analyze the absorption line spectra of six hypothetical stars, each with different temperatures as shown below. For each absorption line spectrum, the short wavelengths of light (blue end) of the EM spectrum are on the left hand side and the long wavelengths of light (red end) of the spectrum are shown on the right side. **Your instructor will assign 3 of the spectra below in this exercise.**

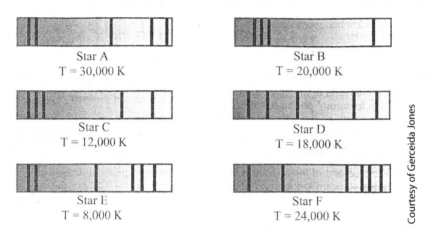

Star A
T = 30,000 K

Star B
T = 20,000 K

Star C
T = 12,000 K

Star D
T = 18,000 K

Star E
T = 8,000 K

Star F
T = 24,000 K

Courtesy of Gerceida Jones

PART III B

In this exercise you will draw spectral curves of the temperature profiles of the Stars from Part IIIA that was assigned by your instructor.

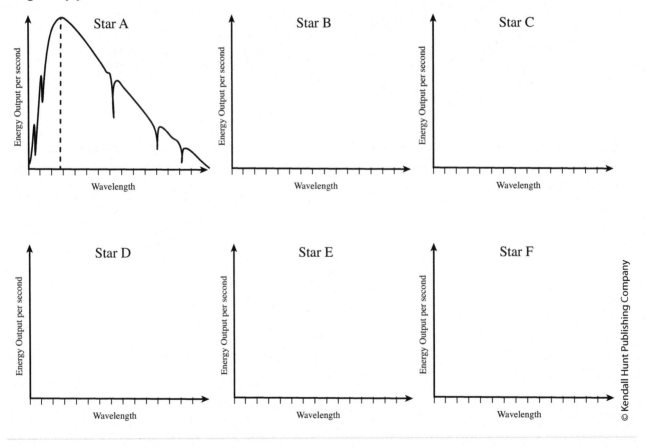

PART IV

Classifying—taking—spectra, and analyzing starlight has led us to the point where we can now plot stars in an organized manner just as an astronomer do. Our final exercise in this lab is to use a H-R diagram **(See Figure 4)** to demonstrate the mass-luminosity relationship of both main sequence—and other—stars which is a chart plotting the luminosities of stars against their surface temperatures. **See Appendix D for additional help. On Page 6–10 you are to plot each star with its number next to it and draw a red line to indicate the main sequence stars just as Figure 4 below.**

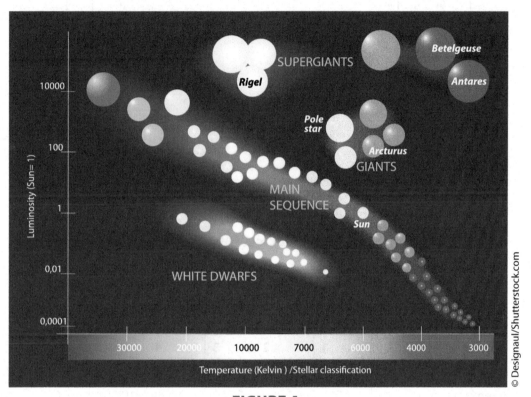

FIGURE 4

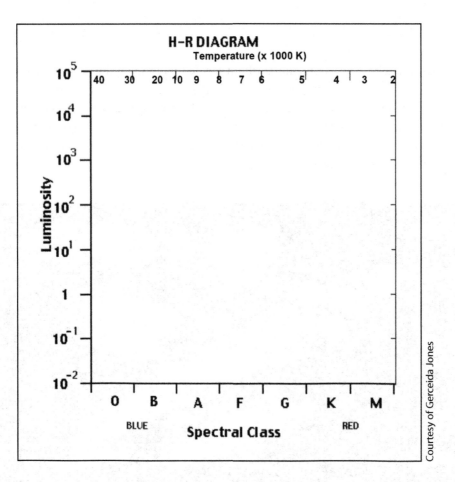

H–R DIAGRAM
Temperature (x 1000 K)

9. a. On H-R diagram above, plot the data for the Main Sequence stars given in the chart. Be sure to label each point with the Star #. Indicate the Main Sequence.

Star #	1	2	3	4	5	6	7
Luminosity	0.05	0.02	1.0	5.0	12	110	8,000
Temperature (K)	3,000	4,000	6,000	8,000	10,000	14,000	16,000

b. On the H-R diagram above, indicate the position of the following star types. Be sure to label each point A-H.

Star Type	Spectral Class	Temperature (K)	Luminosity
A) Main Sequence	G	5,000	5
B) Main Sequence	A	10,000	0.1
C) Main Sequence	A	20,000	10
D) Red Dwarf	M	2,000	0.002
E) Red Giant	M	3,000	5,000
F) Red Super giant	M	4,000	10,000
G) Blue Giant	B	30,000	15,000
H) White Dwarf	A	15,000	0.050

1. Do cold stars always appear to have a different (greater or fewer) number of lines in their absorption spectra than hot stars? Cite evidence from Part IIIA to support your answer.

2. Do cold stars always appear to have more lines at either the blue or red ends of their absorption spectra than hot stars? Cite evidence from Part IIIA to support your answer.

3. Which of these two stars has the highest temperature? _____

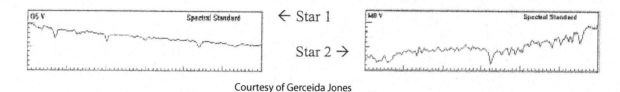

Courtesy of Gerceida Jones

4. Here are the spectra of three stars. Which two have the same spectral type? _____

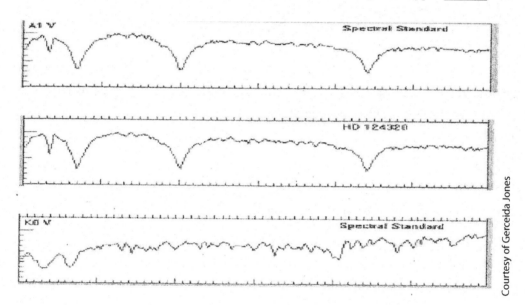

Courtesy of Gerceida Jones

5. Consider the dark line absorption spectra shown below for Star X and Star Z. What can you determine about the relative temperatures of the two stars?

Courtesy of Gerceida Jones

   a. Star X is at the higher temperature.

   b. Star Z is at the higher temperature.

   c. Both stars are the same temperature.

   d. The relative temperatures of the stars cannot be determined.

6. Consider the spectral curves for Star V and Star Y shown in the following graph. What can you determine about the relative temperatures of the two stars?

   a. Star V is at the higher temperature?

   b. Star Y is at the higher temperature?

   c. Both stars are at the same temperature?

   d. The relative temperatures of the stars cannot be determined.

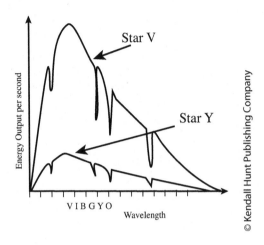

7. The stars Antares and Mimosa each have absolute magnitudes of 4.6. Antares is spectral type M and Mimosa is spectral type B. Which star is larger?

   a. Antares

   b. Mimosa

   c. They are the same size.

   d. There is insufficient information to determine this.

8. Imagine that you are on the surface of Earth (below the atmosphere) and are observing the Sun. Which of the following spectra would you observe by analyzing the sunlight?

   a. dark line absorption spectrum

   b. bright line emission spectrum

   c. continuous spectrum

   d. no spectrum at all

9. Rigel is about 100,000 times as luminous as the Sun and belongs to spectral Class B8. Sirus B is about $^1/_{3000}$ times as luminous as the Sun and belongs to spectral class B8 also. Which star has the greater surface temperature? **Refer to Appendix D, diagram 1.**

   a. Rigel

   b. Sirus B

   c. They have the same temperature.

   d. There is insufficient information to determine this.

10. How does the size of a star near the top left of the H-R diagram compare with a star of the same luminosity near the top right of the H-R diagram?

    a. They are the same size.

    b. The star near the top left is larger.

    c. The star near the top right is larger.

    d. There is insufficient information to determine this.

11. You observe two stars with the same absolute magnitude and determine that one is a spectral type A star while the other is a spectral type F star. Which star has the greater surface area?

    a. the A star

    b. the F star

    c. The surface areas are the same.

    d. There is insufficient information to determine this.

12. How does the Sun produce the energy that heats our planet?

    a. The gases inside the Sun are on fire; they are burning like a giant bonfire.

    b. Hydrogen atoms are combined into helium atoms inside the Sun's core. Small amounts of mass are converted into huge amounts of energy in this process.

    c. When you compress the gas in the Sun, it heats up. This heat radiates outward through the star.

    d. Magnetic energy gets trapped in sunspots and active regions. When this energy is released, it explodes off the Sun as flares that give off tremendous amounts of energy.

    e. The core of the Sun has radioactive materials that give off energy as they decay into other elements.

Appendix A

THE HISTORY AND NATURE OF SPECTRAL CLASSIFICATION

Patterns of absorption lines were first observed in the spectrum of the sun by the German physicist Joseph von Fraunhofer early in the 1800's, but it was not until late in the century that astronomers were able to routinely examine the spectra of stars in large numbers. Astronomers Angelo Secchi and E.C. Pickering were among the first to note that the stellar spectra could be divided into groups by the general appearance of their spectra. In the various classification schemes they proposed, stars were grouped together by the prominence of certain spectral lines. In Secchi's scheme, for instance, stars with very strong hydrogen lines were called type I, stars with strong lines from metallic ions like iron and calcium were called type II, and stars with wide bands of absorption that got darker toward the blue were called type III, and so on. Building upon this early work, astronomers at the Harvard Observatory refined the spectral types and renamed them with letters, A, B, C, etc. They also embarked on a massive project to classify spectra, carried out by a trio of astronomers, Williamina Fleming, Annie Jump Cannon and Antonia Maury. The results of that work, the **Henry Draper Catalog** (named after the benefactor who financed the study), was published between 1918 and 1924, and provided classifications of 225,300 stars. Even this study, however, represents only a tiny fraction of the stars in the sky.

In the course of the Harvard classification study, some of the old spectral types were consolidated together, and the types were rearranged to reflect a steady change in the strengths of representative spectral lines. The order of the spectral classes became O, B, A, F, G, K, and M, and though the letter designations have no meaning other than that imposed on them by history, the names have stuck to this day. Each spectral class is divided into tenths, so that a B0 star follows an O9, and an A0, a B9. In this scheme the sun is designated a type G2.

The early spectral classification system was based on the appearance of the spectra, but the physical reason for these differences in spectra was not understood until the 1930's and 1940's. Then it was realized that, while there were some chemical differences among stars, the main thing that determined the spectral type of a star was its surface temperature. Stars with strong lines of ionized helium (HeII), which were called O stars in the Harvard system, were the hottest, around 40,000 °K, because only at high temperatures would these ions be present in the atmosphere of the star in large enough numbers to produce absorption. The M stars with dark absorption bands, which were produced by molecules, were the coolest, around 3000 °K, since molecules are broken apart (dissociated) at high temperatures. Stars with strong hydrogen lines, the A stars, had intermediate temperatures (around 10,000 °K). The decimal divisions of spectral types followed the same pattern. Thus a B5 star is cooler than a B0 star but hotter than a B9 star.

The spectral type of a star is so fundamental that an astronomer beginning the study of any star will first try to find out its spectral type. If it hasn't already been catalogued (by the Harvard astronomers or the many who followed in their footsteps), or if there is some doubt about the listed classification, then the classification must be done by taking a spectrum of a star and comparing it with an **Atlas** of well-studied spectra of bright stars. Until recently, spectra were classified by taking photographs of the spectra of stars, but modern spectrographs produce digital traces of intensity versus wavelength which are often more convenient to study. *FIGURE A.1* shows some sample digital spectra from the principal spectral types; the range of wavelength (the x axis) is 3900 Å to 4500 Å. The intensity (the y axis) of each spectrum is **normalized**, which means that it has been multiplied by a constant so that the spectrum fits into the picture, with a value of 1.0 for the maximum intensity, and 0 for no light at all.

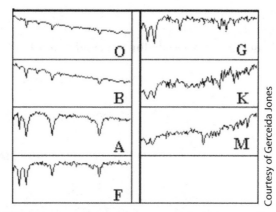

FIGURE A.1   Digital Spectra of the Principal Types

The spectral type of a star allows the astronomer to know not only the temperature of the star, but also its luminosity (expressed often as the absolute magnitude of the star) and its color. These properties, in turn, can help in determining the distance, mass, and many other physical quantities associated with the star, its surrounding environment, and its past history. Thus, knowledge of spectral classification is fundamental to understanding how we put together a description of the nature and evolution of the stars. Looked at on an even broader scale, the classification of stellar spectra is important, as is any classification system, because it enables us to reduce a large sample of diverse individuals to a manageable number of natural groups with similar characteristics. Thus spectral classification is, in many ways, as fundamental to astronomy as is the Linnean system of classifying plants and animals by genus and species.

Since the group members presumably have similar physical characteristics, we can study them as groups, not isolated individuals. By the same token, unusual individuals may readily be identified because of their various differences from the natural groups. These peculiar objects then are subjected to intensive study in order to attempt to understand the reason for their unusual nature. These *exceptions to the rule* often help us to understand broad features of the natural groups. They may even provide evolutionary links between the groups.

Appendix B

How to Use the Telescope

Your instructor will guide you through using the computer program. A quick guide to the commands is listed below. Detailed instructions for the lab are available as an Appendix to the Stellar Spectra Lab on Blackboard.

Quick Program Commands for Spectral Classification of Stars

Start Program
- ► Log in
- ► Type name

Classify Spectra
- ► File
- ► Run
- ► Classify Spectra

Get Unknown Spectrum from List
- ► File
- ► Unknown Spectrum
- ► Program List
- ► Select star

Change Display Options
- ► File
- ► Preferences
- ► Display
- ► Select: intensity trace, photograph, combination

Get Spectral Atlas
- ► File
- ► Atlas of Standard Spectra
- ► Main Sequence

Get Spectral Line table
- ► File
- ► Spectral line table

Open Telescope to take Spectra
- ► File
- ► Return
- ► Exit Classification Window
- ► File
- ► Run
- ► Take Spectra

Using Telescope
- ► Open Dome
- ► Turn on Tracking
- ► Get star in box using E,N,S,W
- ► Change View—get star in crosshairs
- ► Take Reading
- ► Start Count
- ► Save (as Object Number)

Get Saved Spectra
- ► File
- ► Unknown Spectrum
- ► Saved Spectra
- ► Select Spectrum

Appendix C

TABLE I Distinguishing Features of Main Sequence Spectra

SPECTRAL TYPE	SURFACE TEMP (° K)	Distinguishing Features (absorption lines unless noted otherwise)
O	28-40,000	He II lines
B	10-28,000	He I lines; H I Balmer lines in cooler types
A	8-10,000	Strongest H I Balmer at A0; CaII increasing at cooler types; some other ionized metals
F	6000-8000	Ca II stronger; H weaker; Ionized metal lines appearing
G	4900-6000	CaI II strong; Fe and other Metals strong, with neutral metal lines appearing; H weakening
K	3500-4900	Neutral metal lines strong; CH and CN bands developing
M	2000-3500	Very many lines; TiO and other molecular bands; Neutral Calcium prominent. S stars show ZrO and N stars C2 lines as well.
WR (Wolf-Rayet)	40,000+	Broad emission of He II; WC stars show CIII and CIV emission, while WN stars show NII prominently

Appendix D

Life Cycle of an Evolving Star

An H-R diagram is an organized chart showing the stages of a star as it evolves through a life cycle. Main sequence stars, such as our sun, are a range of stars based on size and surface temperature starting from the hot, bright, bluish stars in the upper left hand corner of the diagram to the cool, dim, reddish stars in the lower right hand corner. The life for a new star begins on the main sequence (**See Diagram 1**). As it matures, the star undergoes this remarkable transformation of consuming hydrogen in its core. Once the hydrogen is consumed, the star leaves the main sequence and expands to a red giant. This new stage of development begins the fusion of helium to form heavier elements like Oxygen and Carbon (**See Diagram 2**). This process of expansion, then collapse, then expansion again of the star, forms the light elements that are present in our universe all the way up to Iron.

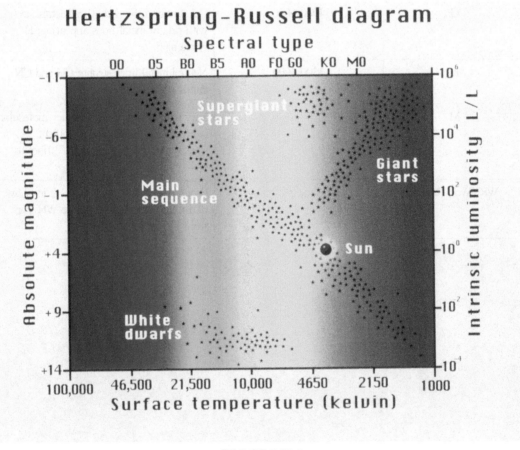

DIAGRAM 1

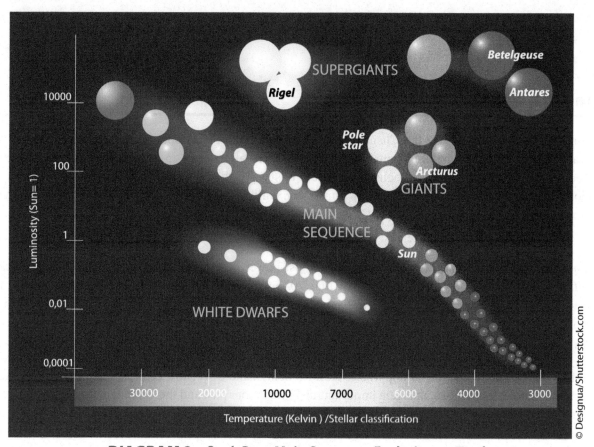

DIAGRAM 2 **Sun's Post-Main Sequence Evolutionary Track**

Appendix E

Chandra X-ray Observatory

Scientists around the world use the Chandra X-ray Observatory over to obtain unprecedented images and spectra of violet, high temperature events and objects that will allow them to better understand the structure and evolution of the Universe. It is a sophisticated piece of state-of-the-art instrumentation that represents a tremendous technological advancement in X-ray astronomy. Unlike the Hubble telescope that orbits Earth every 90 minutes, Chandra circles the Earth every 2.6 days to allow for longer observation periods and to slip beyond Earth's Van Allen radiation belt VARB which causes disruption in images.

The Chandra X-ray Observatory is the world's most powerful X-ray telescope. It has eight-time greater resolution and will be able to detect sources more than 20 times –fainter than any previous X-ray telescope. The Observatory's operating orbit takes it 200-times higher than the Hubble Space Telescope. During each orbit around the Earth, Chandra travels one-third of the way to the moon to allow for the longer observation time and to slip beyond the VARB as previously mentioned.

Why is this X-ray telescope so important? Besides being state-of-the-art technology, scientists can study quasars, black holes, supernova, and gain important insight into our past and the Universe. If you are interested in learning how to decode starlight, then this is the program for you. You will move from pixels to images by choosing your own color scheme in this activity. It will also help you understand how the images they see in the Chandra photo album are generated from the numerical data collected by the observatory. By going to the web site listed below, you will be guided step by step through the data and imaging process with actual data from the Chandra X-ray Observatory. As a student you will develop the data shown in the image, and also, the "false colors" used to display the image. It's exciting! No experience necessary. You want to know how I know? A high school student playing around with the program actually discovered a binary star being formed in another galaxy just by chance. He wrote an article and the rest is history as they say!

http://chandra.harvard.edu/edu/formal/imaging/imaging_high.pdf

Bibliography

1. Bahcall, John N. and Salpeter, Edwin E., "Stellar Energy Generation and Solar Neutrinos", *Physics Today,* October 2005, page 44–45.

2. Kuo, Vince H. and Beichner, Robert J., "Stars of the Big dipper: A 3-D Vector Activity", *The Physics Teacher,* Volume 44, March 2006, page 169, Figure 1.

3. Adams, Jeffrey P., Prather, Edward E., Slater, Timothy F., *Lecture-Tutorials for Introductory Astronomy,* Prentice Hall, Upper Saddle River, NJ 07458, 2005, page 37, Figure 2.

4. Ibid, page 41, temperature profiles, page 6–7.

5. Ibid, pages 43–45 spectral images, pages 6–8 and 6–11.

6. Holland, Arthur and Williams, Mark, "Stellar Evolution: The Life and Death of Our Luminous Neighbors", http://www.umich.edu/~gs265/star.htm, pages 1–2, Figure 4.

7. http://outreach.atnf.csiro.au/education/senior/astrophysics/stellarevolution_hrintro.html

8. http://outreach.atnf.csiro.au/education/senior/astrophysics/stellarevolution_deathlow.html

LAB 6

Hubble's Law

This lab was adapted with permission from *The Hubble Redshift Distance Relation* lab: A Manual to Accompany Software for the Introductory Astronomy Lab Exercise Document SM 3: Version 1.

OBJECTIVES

- ▶ To understand Hubble's law and its implication for the expansion of the universe
- ▶ To use a simulated telescope and spectrometer to collect spectra and apparent magnitudes of galaxies
- ▶ To calculate the distance and velocity of galaxies and use them to determine Hubble's parameter and the age of the universe

INTRODUCTION

The late biologist J.B.S. Haldane once wrote: "The universe is not only queerer than we suppose, but queerer than we can suppose." One of the oddest things about the universe is that virtually all the galaxies in it (with the exception of a few nearby ones) are moving away from the Milky Way. This curious fact was first discovered in the early 20th century by astronomer Vesto Slipher, who noted that absorption lines in the spectra of most spiral galaxies had longer wavelengths, or more red shifted, than those observed from stationary objects. Assuming that the red shift was caused by the Doppler shift, Slipher concluded that the red shifted galaxies were all moving away from us.

In the 1920's, Edwin Hubble measured the distances of the galaxies for the first time, and when he plotted these distances against the velocities for each galaxy he noted something even odder: The further a galaxy was from the Milky Way, the faster it was moving away. Was there something special about our place in the universe that made us a center of cosmic repulsion? He summarized this relationship into a formula: $V = H_0 \times D$. H_0 is Hubble's parameter and helps determine how fast the galaxies are moving away from one another.

Astrophysicists readily interpreted Hubble's relation as evidence of a universal expansion. The distance between all galaxies in the universe was getting bigger with time, like the distance between raisins in a rising loaf of bread. An observer on *ANY* galaxy, not just our own, would see all the other galaxies traveling away, with the furthest galaxies traveling the fastest.

This was a remarkable discovery. The expansion is believed today to be a result of a "Big Bang" which occurred between 10 and 20 billion years ago, a date which we can calculate by making measurements like those of Hubble. *The rate of expansion of the universe tells us how long it has been expanding.* The universe is continuing to expand as we now know from the winners of the 2011 Nobel Prize in Physics, the expansion is accelerating according to white dwarf supernova data. We determine the rate by plotting the **velocities of galaxies against their distances**, and determining the slope of the graph, a number called the Hubble Parameter, H_o, which tells us how fast a galaxy at a given distance is receding from us. Hubble's discovery of the correlation between velocity and distance is fundamental in reckoning the history of the universe.

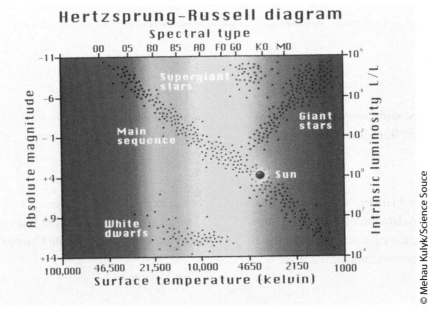

FIGURE 1   Hubble's graph

Using modern techniques of digital astronomy, we will repeat Hubble's experiment. The technique we will use is fundamental to cosmological research these days. Even though Hubble's first measurements were made three-quarters of a century ago, we have still only measured the velocities and distances of a small fraction of the galaxies we can see, and so we have only small amount of data on whether the rate of expansion is the same in all places and in all directions in the universe. The red shift distance relation thus continues to help us map the universe in space and time.

Measuring Distance to Galaxies

One of the measurements used to determine Hubble's parameter is distance. Measuring the distance to a galaxy poses a problem since they are so far away that the methods we use on earth, i.e. rulers will not work. It is also made more complicated because when we view a galaxy we have to determine is it faint because it is very small or because it is very far away? To help us determine the distance to a galaxy we need to determine how bright the galaxy appears (**apparent magnitude**) and compare it to how much light it is actually radiating (**absolute magnitude**). Apparent magnitude is related to distance by the inverse square law. In other words, light intensity decreases as distance is squared. A light source that is three times the distance from an observer appears to be nine times as faint as a close source. Absolute magnitude is defined by being the apparent magnitude measured at a standard distance of 10 parsecs. If you could put all the objects at the same distance from the observer, you could cancel out the effect of distance and be able to determine the actual amount of light radiating from a galaxy. There is a special relationship between these two values that we will use to calculate distance. The difference between the apparent magnitude (m) and the absolute magnitude (M) is called the **distance modulus (m – M)**. As a formula:

$$m - M = 5 \log (\text{distance in parsecs}/10)$$

Isolating distance we get:

$$D = 10^{(m - M + 5)/5}$$

This is the formula we will use in the lab today to measure distance in parsecs. For simplicity during this lab we will assume that the absolute magnitude (M) of each galaxy is –21. Of course in a real situation you would measure the distance to a galaxy using Cepheid variable stars or supernovae, then measure the apparent magnitude and scale it to a standard distance of 10 parsecs using the inverse square relationship. This would then give you the absolute magnitude of your galaxy. You will actually collect the apparent magnitude (m) today using the spectrometer attached to the telescope.

Sample Calculation 1—Determining Distance

Formula: $D = 10^{(m - M + 5)/5}$

D = distance to galaxy in parsecs

m = apparent magnitude = 16.87

Constant — M = absolute magnitude = - 21

▶ $D = 10^{(16.87 - (-21) + 5)/5} = 10^{(8.574)} = 3.750 \times 10^8$ parsecs

(Make sure you get this answer on your calculator; different calculators require different steps to solve $10^{(8.574)}$ which equal 3.750×10^8)

▶ This needs to be converted to Megaparsecs.

$1 \text{ Mpc} = 1.0 \times 10^6 \text{ pc}$

$D = 3.750 \times 10^8 \text{ pc} \times 1 \text{ Mpc}/1 \times 10^6 \text{ pc}$

$= 3.750 \times 10^2 \text{ Mpc} \ (375.0 \text{ Mpc})$

Measuring the Velocity of Galaxies

The velocity of a galaxy can be calculated by looking at its absorption spectrum. You should remember from lecture and lab that each element produces a specific pattern of wavelengths (λ) when it absorbs energy (see Atomic Spectroscopy lab to review). You should also remember that an object in motion will exhibit a Doppler shift in its spectrum, either towards the red end of the spectrum or the blue end. Objects moving toward us exhibit a blue shift and objects moving away exhibit a red shift. Since we know that each element shows a specific pattern of wavelengths we can use this for reference lines to compare to lines in the shifted spectrum. We can use this information to come up with a formula to determine the velocity of the moving galaxy. For today's lab we will use as a reference line the Calcium K line which is located at 3933.67 Ångstroms. The formula is below:

$$\textbf{Velocity}_{\text{galaxy}} = (\lambda_{\text{kmeasured}} - \lambda_{\text{kreference}} / \lambda_{\text{reference}}) \times \textbf{c}$$

Where c is the speed of light = 3.0×10^5 km/s

Sample Calculation 2—Determining Velocity

Formula: $V = (\lambda_{\text{kmeasured}} - \lambda_{\text{kreference}} / \lambda_{\text{reference}}) \times c$

V = velocity of galaxy

$\lambda_{\text{kmeasured}} = 4562$ Å

Constant $\lambda_{\text{kreference}} = 3933.67$ Å

$c = 3.0 \times 10^8$ m/s $= 3 \times 10^5$ km/s

$V = (4562$ Å $- 3933.67$ Å $/ 3933.67$ Å$) \times 3.0 \times 10^5$ km/s

$= 4.792 \times 10^4$ km/s

Determining the Age of the Universe

You have learned that there is a linear relationship between the distance of a galaxy and its recessional velocity that is represented by the equation $v = H_o \times D$. You have also learned that there is a specific relationship between velocity, distance and time of any object in motion, $v = D/t$. Since galaxies are in motion we can integrate these two equations to come up with an equation that tells us something about time. The galaxies have been traveling since the Big Bang, the beginning of the universe, and if we can calculate how long they have been traveling then we essentially know the age of the universe.

$$V = \boxed{(H_o \times D)} \qquad t = D/v \qquad t = D/H_o \times D \qquad t = 1/H_o$$

So to calculate the age of the universe all you need is Hubble's parameter. Now some conversions need to be done because we want time (or age) in years and the unit for H_o is km/s/Mpc.

Sample Calculation 3—Determining the Age of the Universe

Formula: $t = 1/H_o$

$H_o = 75$ km/s/Mpc

▶ We need to "get rid of" the distance units (km and Mpc) since we are solving for time (s). Here are some conversion units we can use:

1 Mpc $= 3.25 \times 10^6$ light years   1 light year $= 1.0 \times 10^{13}$ km.

▶ We can do this in one step:

$t = 1/75$ km/s/Mpc $\times (3.25 \times 10^6$ lt yr / 1 Mpc$) \times (1 \times 10^{13}$ km / 1 lt yr$)$

$= 4.33 \times 10^{17}$ s

(Notice how all units cancel out except seconds)

▶ Now all we have to do is convert seconds to years.

1 year $= 3.15 \times 10^7$ s

$t = 4.33 \times 10^{17}$ s $\times (1$ yr / 3.15×10^7 s$)$

$= 1.37 \times 10^{10}$ years or 13.7 billion years.

Does this make sense? Yes, using other methods the age of the universe is estimated to be around 14 billion years old.

During the Lab

We will simulate an evening's observation during which we will collect data and draw conclusions on the rate of expansion of the universe. You will collect data on **thirteen** different galaxies. We will work on the first galaxy together and you will collect the remaining twelve on your own. You will notice that the telescope controls and data collection are very similar to the Stellar Spectroscopy lab you did earlier.

1. Open the **Hubble Redshift** program by double-clicking on the icon. Click **File** and then **Log in...** in the MENU BAR and enter your name. Click **ok** when ready.

2. The title screen appears. In the MENU BAR select **Telescope**. The software simulates the operation of a computer-controlled spectrometer attached to a telescope at a large mountaintop observatory. The screen shows the control panel and view window as found in the "warm room" at the observatory.

3. To begin, first open the dome by clicking on the **dome** button. The dome opens and the view we see is from the finder scope. The finder scope is mounted on the side of the main telescope and points in the same direction. You can change the view from finder scope to main telescope by clicking on **telescope**. Notice the stars are drifting in the view window. This is due to the rotation of the earth and is noticeable under the high magnification of the finder telescope. To keep an object centered we need to turn on the **tracking** of the telescope by clicking on it.

4. To collect data you need to do the following:

   a. Click on **Turn on Control Panel**

   b. Select **Tools**.

   c. Select **SLEW, then Observational Hot List.**

   d. Start with item 1 in the hot list up to item 13 recording the data.

   e. Example: Choose item #1, the telescope will pan to the correct galaxy; write the galaxy name and the apparent magnitude in data table 1.

   f. Using the **Spectrum Measuring Tool,** measure the Ca K line and record the results in Data Table 1. **Save** the data. The program will automatically save in the galaxy's name.

   g. Continue to collect the data for all 13 galaxies.

5. To switch to the data collection screen, click the **take reading** button. A new window will open. We will be looking at the spectrum of the galaxy as collected through the slit of the spectrometer. The spectrum will include the characteristic H and K Calcium lines (we are only concerned with the K lines) that would normally appear at 3968.847 Å and 3933.67 Å respectively if the galaxies were not moving. However, the H and K lines will be red shifted to longer wavelengths depending on how fast the galaxy is receding.

6. To initiate data collection, click on **start/resume count** in the MENU BAR. After 10 or so seconds hit **stop count** so that you can observe and collect data from the spectrum. In **Data Table 1** you should record the following information:

a. Object name

b. Apparent magnitude

c. Wavelength of the K line (left clicks on the peak farthest to the left and position the red crosshair to the center of the peak). Wavelength will appear at the top.

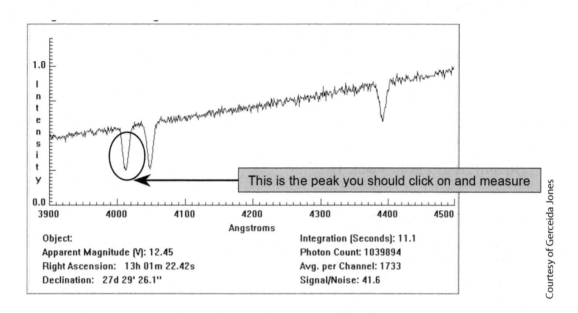

Object:
Apparent Magnitude (V): 12.45
Right Ascension: 13h 01m 22.42s
Declination: 27d 29' 26.1"

Integration (Seconds): 11.1
Photon Count: 1039894
Avg. per Channel: 1733
Signal/Noise: 41.6

Courtesy of Gerceida Jones

FIGURE 2   Finding Calcium K line

To collect data for the additional galaxies, press **return** and then **ok** and then **yes** to return to the telescope window. **Change the view** to finder and follow steps 4–8. You need to collect **one galaxy from 5 different star fields** for a total of 5 galaxies. (Don't use item #7, it is too far away to get any good readings). Remember to **record your data in Data Table 1**.

Calculations

[handwritten notes at top right: m - Apparent magnitude / M: Absolute magnitude. / -21 for absolute magnitude. / Convert everything into whole numbers]

1. To calculate **distance**s in parsecs (pc) see sample calculation 1.

   a. $D = 10^{(m - M + 5)/5}$

   Remember that M is assumed to be –21 for all galaxies

   b. Convert distance to Megaparsecs see sample calculation 1.

   (Record distance in data table 1)

2. To calculate **velocities in km/s see** sample calculation 2.

   $$V_K = (\lambda_{Kmeasured} - \lambda_{Kreference} / \lambda_{Kreference}) \times c$$

   Remember $\lambda_{kreference}$ = 3933.67 Ångstroms

   c = speed of light = 3×10^5 km/s

   (Record velocity in data table 1)

3. To find **Hubble's parameter H_o**:

   a. Use *Excel* to plot distance (Mpc) vs. velocity (km/s). Refer to your handbook Appendix 2 and lab skills handouts to refresh your memory.

   b. Have Excel draw a **linear trend line**, set **intercept = 0** and **display the equation** on the graph.

   c. The **slope** of the line is Hubble's parameter, H_o.

   (Hint: it should be between 50 and 100 km/s/Mpc).

   (Record H_o in Data Table 2)

4. To calculate the **Age of the Universe** see sample calculation 3.

   a. $t = 1/H_o$

   b. 1 Mpc = 3.25×10^6 light years and 1 light year = 1×10^{13} km

   (Record age of the universe in Data Table 2)

****You MUST show all work and attach your graph to receive full credit!**

Additional Sources:

Moché, Dinah. (2002) *Astronomy: A Self-Teaching Guide.* John Wiley and Sons. New York.

Name _____

DATA TABLE 1—Distance and Velocity

Hot List RA/Dec	Galaxy Name (Object)	Apparent Magnitude (m)	$\lambda_{measured}$ Calcium K line (Å)	Distance in Mpc (D)	Velocity in Km/s (V_k)
1 11h46m51.80s 55°42'18.0"	36747	15.60	4130.00		
2 11h47m45.60s 55°41'24.0"	36773	15.8	4126.67		
3 11h47m45.60s 55°46'22.0"	36805	15.90	4133.33		
4 12h59m35.13s 27°57'36.0"	NGC4874	12.9	4026.67		
5 13h00m08.27s 27°58'40.0"	NGC4889	12.5	4019.33		
6 14h32m44.27s 31°33'48.0"	51976	17.90	4446.67		
7 14h32m44.40s 31°35'54.0"	51975	18.50	44 88.33		
8 15h22m24.60s 27°43'21.0"	54876	16.00	4205.00		
9 15h22m24.60s 27°44'27.0"	54875	16.90	4243.33		
10 15h22m42.53s 27°40'22.0"	54891	16.30	4223.33		
11 23h10m22.50s 7°34'54.0"	NGC7499	14.10	4085.00		
12 23h10m22.50s 7°35'22.0"	NGC7501	14.70	4100.00		
13 23h10m42.33s 7°34'01.5"	NGC7503	14.90	4105.00		

Hubbles
constant
should be
between
500
100

DATA TABLE 2—Calculating H_0 and Age of the Universe

Hubble's parameter (H_0)	Km/s/Mpc
Age of universe	Years

****You MUST show all work and attach your graph to receive full credit!

(ALL QUESTIONS MUST BE TYPED, EXCEPT MATH)

1. If the Hubble parameter were to be found to be much smaller than we think it is, how would this change the measured age of the universe? Justify your answer.

2. Using Hubble's law, determine the recessional velocity of a galaxy 700 Mpc away. Use 75 km/s/Mpc for your Hubble's parameter. Show your work for full credit.

3. If a galaxy has the apparent magnitude of 13.97, what is its distance from the Milky Way? Assume an absolute magnitude (M) of –21.

4. Some quasars (quasi-stellar radio source) have the largest red shifts ever observed. If this phenomenon is due to a Doppler shift, what can you say about their distance? What can you predict about their age compared to the universe?

Note: The fact that the recession rates are the same in all directions is the condition called isotropy. The Universe is isotropic.

LAB 7

Visit to the American Museum of Natural History

Take this page, a pen, and notebook with you when you make your visit.

OBJECTIVES

▶ To learn about the formation and evolution of the universe.
▶ To learn why the Earth is habitable.

INTRODUCTION

For next week's lab, you will visit the American Museum of Natural History with the tour or on your own. The activity might be more enjoyable if you take your friends from class with you! **This lab's due date TBA.**

The American Museum of Natural History is located on Central Park West between W. 77th and W. 81st streets. Take the **C** train (W. 3rd St. and 6th Ave.) to W. 81st street. The museum is open daily, 10:00 a.m.–5:45 p.m. Admission charge is voluntary. **Pay what you wish.** You should bring your NYU student ID to show that you are completing an assignment for class. You may do as a group lab if you so choose.

KEEP YOUR RECEIPT(S). Staple it to your answers to turn in.

You will receive NO CREDIT (0) without a dated ticket. You must get a ticket. If it is a group lab, everyone must staple their receipt to the lab with name written on the ticket. Please don't get there after 4 pm since you will not be able to get a receipt. You may work in groups. Thank you.

BACKGROUND

Before you visit the museum, go to the AMNH website (http:/www.AMNH.org) to learn about the museum.

DURING THE LAB

(Staple your receipt(s) to your answers that you turn in.)

All answers must be typed. ANSWERS SHOULD BE IN YOUR OWN WORDS. DO NOT REHASH WHAT IS WRITTEN ON THE MUSEUM DISPLAYS. OTHERWISE IT IS CONSIDERED PLAGIARISM.

PART I

Dorothy and Lewis B. Cullman Hall of the Universe (Lower Level, Rose Center of Earth and Space, Hayden Planetarium)

The Cullman Hall of the Universe sits under the spherical Space Theater and is structured around four metallic, interactive panels labeled: Universe, Galaxies, Stars, and Planets. Walk and browse the four panels and stop at the Universe Panel. Read the placard titled "Formation and Evolution of the Universe" next to the coronagraph responsible for detecting the first brown dwarf courtesy of Ben R. Oppenheimer. This panel briefly summarizes the origins of the universe.

A. Universe

1. How do astronomers use light to observe the evolution of the universe? How do astronomers use light to support the Big Bang Theory?

B. Galaxies

1. Read the first panel titled "GALAXIES: in the Universe". Roughly how many galaxies are in our Local Group? What force binds them together? Look into the ovular window titled "The Local Group". Name 6 galaxies that are a part of our Local Group. Which 2 galaxies in our Local Group are the largest?
2. Name and describe three types of galaxies that reside in our universe.

C. Stars

1. What star is nearest our Sun? What star is the brightest in our nighttime sky?
2. Focusing on the following groups of stars: low mass, intermediate mass, high mass, and very high mass, describe the "death" of these stars and the remnants produced. You **MUST** provide your answer in a **chart format.**
3. Explain how stars are made, in part, from the ashes of previous generations.

D. Planets

1. In our planetary system, what are the 5 classes of objects that orbit our Sun? Where does Pluto reside?
2. Choose 2 terrestrial planets in our solar system you would NOT be comfortable living on. Give one reason for each.

WARNING: If this exhibit is still closed, you know enough info from lecture and how to research these questions on your own. **Complete all of Part II.**

PART II

David S. and Ruth L. Gottesman Hall of Planet Earth

Head upstairs to the Gottesman Hall of Planet Earth to learn what makes Earth different from its fellow terrestrial planets and how those differences make it possible for life on Earth. Facing the **Planets Panel**, turn left and head up the curved ramp. Take the escalators straight ahead to the First Floor. The Gottesman Hall of Planet Earth will be directly behind you.

1. Why is Earth able to support life? Provide evidence based on the following: protective shield, proper temperature range, H_2O, and right "ingredients" (be sure to include what they are).
2. How can deep-sea vent organisms survive without sunlight and photosynthesis? What is the significance of this discovery (provide three lines of evidence)? What can this say about our search for extraterrestrial life?
3. How do we know the age of the Earth? How do we know about the Earth's early atmosphere? Refer to the banded iron formation in your explanation.

CPSIA information can be obtained
at www.ICGtesting.com
Printed in the USA
LVOW02s0844100917
548000LV00002B/5/P